Lecture Notes in Computer Science 4446

Commenced Publication in 1973
Founding and Former Series Editors:
Gerhard Goos, Juris Hartmanis, and Jan van Leeuwen

T0223236

Carlos Cotta   Jano van Hemert (Eds.)

Evolutionary Computation in Combinatorial Optimization

7th European Conference, EvoCOP 2007
Valencia, Spain, April 11-13, 2007
Proceedings

 Springer

Volume Editors

Carlos Cotta
Universidad de Málaga, Dept. Lenguajes y Ciencias de la Computación
ETSI Informática, Campus Teatinos, 29071 Málaga, Spain
E-mail: ccottap@lcc.uma.es

Jano van Hemert
University of Edinburgh, National e-Science Institute
15 South College Street, Edinburgh EH8 9AA, UK
E-mail: jano@vanhemert.co.uk

Cover illustration: Morphogenesis series #12 by Jon McCormack, 2006

Library of Congress Control Number: 2007923599

CR Subject Classification (1998): F.1, F.2, G.1.6, G.2.1, G.1

LNCS Sublibrary: SL 1 – Theoretical Computer Science and General Issues

ISSN   0302-9743
ISBN-10   3-540-71614-9 Springer Berlin Heidelberg New York
ISBN-13   978-3-540-71614-3 Springer Berlin Heidelberg New York

Springer is a part of Springer Science+Business Media

springer.com

© Springer-Verlag Berlin Heidelberg 2007
Printed in Germany

Typesetting: Camera-ready by author, data conversion by Scientific Publishing Services, Chennai, India
Printed on acid-free paper   SPIN: 12041480   06/3180   5 4 3 2 1 0

Preface

Metaheuristics have often been shown to be effective for difficult combinatorial optimization problems appearing in various industrial, economical, and scientific domains. Prominent examples of metaheuristics are evolutionary algorithms, simulated annealing, tabu search, scatter search, memetic algorithms, variable neighborhood search, iterated local search, greedy randomized adaptive search procedures, estimation of distribution algorithms, and ant colony optimization. Successfully solved problems include scheduling, timetabling, network design, transportation and distribution, vehicle routing, the traveling salesman problem, satisfiability, packing and cutting, and general mixed integer programming.

EvoCOP began in 2001 and has been held annually since then. It was the first event specifically dedicated to the application of evolutionary computation and related methods to combinatorial optimization problems. Originally held as a workshop, EvoCOP became a conference in 2004. The events gave researchers an excellent opportunity to present their latest research and to discuss current developments and applications as well as providing for improved interaction between members of this scientific community. Following the general trend of hybrid metaheuristics and diminishing boundaries between the different classes of metaheuristics, EvoCOP has broadened its scope over the last years and invited submissions on any kind of metaheuristic for combinatorial optimization.

This volume contains the proceedings of EvoCOP 2007, the seventh European Conference on Evolutionary Computation in Combinatorial Optimization. It was held in Valencia, Spain, April 11–13, 2007, jointly with EuroGP 2007, the Tenth European Conference on Genetic Programming, EvoBIO 2007, the Fifth European Conference on Evolutionary Computation and Machine Learning in Bioinformatics, and EvoWorkshops 2007, which consisted of the following seven individual workshops: EvoCOMNET, the Fourth European Workshop on the Application of Nature-Inspired Techniques to Telecommunication Networks and Other Connected Systems; EvoFIN, the First European Workshop on Evolutionary Computation in Finance and Economics; EvoIASP, the Ninth European Workshop on Evolutionary Computation in Image Analysis and Signal Processing; EvoInteraction, the Second European Workshop on Interactive Evolution and Humanized Computational Intelligence; EvoMUSART, the Fifth European Workshop on Evolutionary Music and Art; EvoSTOC, the Fourth European Workshop on Evolutionary Algorithms in Stochastic and Dynamic Environments, and EvoTransLog, the First European Workshop on Evolutionary Computation in Transportation and Logistics. Since 2007, all these events are grouped under the collective name EvoStar, and constitute Europe's premier co-located meetings on evolutionary computation.

Accepted papers of previous EvoCOP editions were published by Springer in the series *Lecture Notes in Computer Science* (LNCS – Volumes 2037, 2279, 2611, 3004, 3448, and 3906).

EvoCOP	Submitted	Accepted	Acceptance ratio
2001	31	23	74.2%
2002	32	18	56.3%
2003	39	19	48.7%
2004	86	23	26.7%
2005	66	24	36.4%
2006	77	24	31.2%
2007	81	21	25.9%

The rigorous, double-blind reviewing process of EvoCOP 2007 resulted in a strong selection among the submitted papers; the acceptance rate was 25.9%. Each paper was reviewed by at least three members of the International Program Committee. All accepted papers were presented orally at the conference and are included in this proceedings volume. We would like to credit the members of our Program Committee, to whom we are very grateful for their quick and thorough work and the valuable advice on how to improve papers for the final publication. EvoCOP 2007 contributions deal with representations, heuristics, analysis of problem structures, and comparisons of algorithms. The list of studied combinatorial optimization problems includes prominent examples like graph coloring, knapsack problems, the traveling salesperson problem, scheduling, as well as specific real-world problems.

We would like to express our sincere gratitude to the internationally renowned invited speakers who gave the keynote talks at the conference: Ricard V. Solé, head of the Complex Systems Lab at the University Pompeu Fabra, Chris Adami, head of the Digital Life Lab at the California Institute of Technology, and Alan Bundy, from the Centre for Intelligent Systems and their Applications, School of Informatics at the University of Edinburgh.

The success of the conference resulted from the input of many people, to whom we would like to express our appreciation. We thank Marc Schoenauer for providing the Web-based conference management system. The local organizers, led by Anna Isabel Esparcia-Alcázar, did an extraordinary job for which we are very grateful. We thank the Universidad Politécnica de Valencia, Spain, for their institutional and financial support and for providing premises and administrative assistance, the Instituto Tecnológico de Informática in Valencia for cooperation and help with local arrangements, and the Spanish Ministerio de Educación y Ciencia for their financial support. Thanks are also due to Jennifer Willies and the Centre for Emergent Computing at Napier University in Edinburgh, Scotland, for administrative support and event coordination. Last, but not least, we would especially like to thank Jens Gottlieb and Günther Raidl for their support and guidance, to whom we owe a lot. From their hard work and dedication, EvoCOP 2007 has now become one of the reference events in evolutionary computation.

April 2007   Carlos Cotta
                                                                 Jano van Hemert

Organization

EvoCOP 2007 was organized jointly with EuroGP 2007, EvoBIO 2007, and EvoWorkshops 2007.

Organizing Committee

Chairs   Carlos Cotta, Universidad de Málaga, Spain,

                     Jano van Hemert, University of Edinburgh, UK

Local Chair   Anna Isabel Esparcia-Alcázar, Universidad Politécnica de
                     Valencia, Spain

Publicity Chair   Leonardo Vanneschi, University of Milano-Bicocca, Italy

EvoCOP Steering Committee

Carlos Cotta, Universidad de Málaga, Spain,

Jens Gottlieb, SAP AG, Germany,

Jano van Hemert, University of Edinburgh, UK,

Günther Raidl, Vienna University of Technology, Austria

Program Committee

Adnan Acan, Eastern Mediterranean University, Turkey
Hernán Aguirre, Shinshu University, Japan
Enrique Alba, Universidad de Málaga, Spain
Mehmet Emin Aydin, London South Bank University, UK
Ruibin Bai, University of Nottingham, UK
Christian Bierwirth, University of Bremen, Germany
Christian Blum, Universitat Politècnica de Catalunya, Spain
Peter Brucker, University of Osnabrück, Germany
Edmund Burke, University of Nottingham, UK
Pedro Castillo, Universidad de Granada, Spain
David W. Corne, Heriot-Watt University, UK
Ernesto Costa, University of Coimbra, Portugal
Carlos Cotta, Universidad de Málaga, Spain
Peter I. Cowling, University of Bradford, UK
Bart Craenen, Napier University, Edinburgh, UK
Keshav Dahal, University of Bradford, UK
David Davis, NuTech Solutions Inc., USA
Karl F. Dörner, University of Vienna, Austria

Table of Contents

A New Local Search Algorithm for the DNA Fragment Assembly Problem

Enrique Alba and Gabriel Luque

Grupo GISUM, Departamento de LCC
E.T.S.I. Informática
Campus Teatinos, 29071 Málaga (Spain)
{eat,gabriel}@lcc.uma.es

Abstract. In this paper we propose and study the behavior of a new heuristic algorithm for the DNA fragment assembly problem: PALS. The DNA fragment assembly is a problem to be solved in the early phases of the genome project and thus is very important since the other steps depend on its accuracy. This is an NP-hard combinatorial optimization problem which is growing in importance and complexity as more research centers become involved on sequencing new genomes. Various heuristics, including genetic algorithms, have been designed for solving the fragment assembly problem, but since this problem is a crucial part of any sequencing project, better assemblers are needed. Our proposal is a very efficient assembler that allows to find optimal solutions for large instances of this problem, considerably faster than its competitors and with high accuracy.

1 Introduction

With the advance of computational science, bioinformatics has become more and more attractive to researchers in the field of computational biology. Genomic data analysis using computational approaches is very popular as well. The primary goal of a genomic project is to determine the complete sequence of the genome and its genetic contents. Thus, a genome project is accomplished in two steps, the first one is the genome sequencing and the second one is the genome annotation (i.e., the process of identifying the boundaries between genes and other features in raw DNA sequence).

In this paper, we focus on the genome sequencing, which is also known as the DNA fragment assembly problem. The fragment assembly occurs in the very beginning of the process and therefor other steps depend on its accuracy. At present, DNA sequences that are longer than 600 base-pairs (bps) cannot routinely be sequenced accurately. For example, human DNA is about 3.2 billion nucleotides in length and cannot be read at once. Hence, large strands of DNA need to be broken into small fragments for sequencing in a process called *shotgun sequencing*. In this approach, several copies of a portion of DNA are each broken into many segments short enough to be sequenced automatically by machine. But this process does not keep neither the ordering of the fragments nor the

C. Cotta and J. van Hemert (Eds.): EvoCOP 2007, LNCS 4446, pp. 1–12, 2007.
© Springer-Verlag Berlin Heidelberg 2007

portion from which a particular fragment came. This leads to the DNA fragment assembly problem [1] in which these short sequences have to be reassembled to their (supposed) original form. The automation allows shotgun sequencing to proceed far faster than traditional methods. But comparing all the tiny pieces and matching up the overlaps requires massive computation.

The assembly problem is therefore a combinatorial optimization problem that, even in the absence of noise, is NP-hard: given k fragments, there are $2^k k!$ possible combinations. Over the past decade a number of fragment assembly packages have been developed and used to sequence different organisms. The most popular packages are PHRAP [2], TIGR assembler [3], STROLL [4], CAP3 [5], Celera assembler [6], and EULER [7]. These packages deal with the previously described challenges to different extents, but none of them solves all of them. Each package automates fragment assembly using a variety of algorithms. The most popular techniques are greedy-based while other approaches have tackled the problem with metaheuristics [8]. This work reports on the design and implementation of a new problem aware local search algorithm to find fast and accurate solutions for large instances of the DNA fragment assembly problem. We additionally study the behavior of several variants of the basic method. Finally, we also compare the results of our approach with the ones of classical (real world) assemblers in order to test the actual interest of our method.

The remainder of this paper is organized as follows. In the next section, we present background information about the DNA fragment assembly problem. In Section 3, the details of our proposed heuristic are presented. We analyze the results of our experiments in Section 4. Finally, we end this paper by giving our final thoughts and conclusions in Section 5.

2   The DNA Fragment Assembly Problem

In order to determine the function of specific genes, scientists have learned to read the sequence of nucleotides comprising a DNA sequence in a process called DNA sequencing. To do that, multiple exact copies of the original DNA sequence are made. Each copy is then cut into short fragments at random positions. These are the first three steps depicted in Fig. 1 and they take place in the laboratory. After the fragment set is obtained, a traditional assemble approach is followed in this order: overlap, layout, and then consensus. To ensure that enough fragments overlap, the reading of fragments continues until a coverage is satisfied. These steps are the last three ones in Fig. 1. In what follows, we give a brief description of each of the three phases, namely overlap, layout, and consensus.

Overlap Phase - Finding the overlapping fragments. This phase consists of finding the best or longest match between the suffix of one sequence and the prefix of another. In this step, we compare all possible pairs of fragments to determine their similarity. Usually, a dynamic programming algorithm applied to semiglobal alignment is used in this step. The intuition behind finding the pairwise overlap is that fragments with a significant overlap score are very likely next to each other in the target sequence.

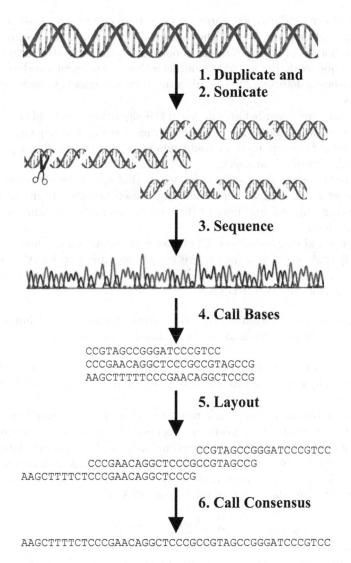

Fig. 1. Graphical representation of DNA sequencing and assembly

Layout Phase - Finding the order of fragments based on the computed similarity score. This is the most difficult step because it is hard to tell the true overlap due to the following challenges:

1. Unknown orientation: After the original sequence is cut into many fragments, the orientation is lost. One does not know which strand should be selected. If one fragment does not have any overlap with another, it is still possible that its reverse complement might have such an overlap.

2. Base call errors: There are three types of base call errors: substitution, insertion, and deletion errors. They occur due to experimental errors in the electrophoresis procedure (the method used in the laboratories to read the ADN sequences). Errors affect the detection of fragment overlaps. Hence, the consensus determination requires multiple alignments in highly coverage regions.
3. Incomplete coverage: It happens when the algorithm is not able to assemble a given set of fragments into a single contig. A contig is a sequence in which the overlap between adjacent fragments is greater or equal to a predefined threshold (cutoff parameter).
4. Repeated regions: "Repeats" are sequences that appear two or more times in the target DNA. Repeated regions have caused problems in many genome-sequencing projects, and none of the current assembly programs can handle them perfectly.
5. Chimeras and contamination: Chimeras arise when two fragments that are not adjacent or overlapping on the target molecule join together into one fragment. Contamination occurs due to the incomplete purification of the fragment from the vector DNA.

After the order is determined, the progressive alignment algorithm is applied to combine all the pairwise alignments obtained in the overlap phase.

Consensus Phase - Deriving the DNA sequence from the layout. The most common technique used in this phase is to apply the majority rule in building the consensus.

To measure the quality of a consensus, we can look at the distribution of the coverage. Coverage at a base position is defined as the number of fragments at that position. It is a measure of the redundancy of the fragment data, and it denotes the number of fragments, on average, in which a given nucleotide in the target DNA is expected to appear. It is computed as the number of bases read from fragments over the length of the target DNA [1].

$$Coverage = \frac{\sum_{i=1}^{n} length\ of\ the\ fragment\ i}{target\ sequence\ length} \tag{1}$$

where n is the number of fragments. The higher the coverage, the fewer number of the gaps, and the better the result.

3 Our Proposal: Problem Aware Local Search (PALS)

Classical assemblers use fitness functions that favor solutions in which strong overlap occurs between adjacent fragments in the layouts, using equations like 2 [9] (where $w_{i,j}$ is the overlap between fragments i and j). But the actual objective is to obtain an order of the fragments that minimizes the number of contigs, with the goal of reaching one single contig, i.e., a complete DNA sequence composed of all the overlapping fragments. Therefore, the number of contigs is used as a high-level criterion to judge the whole quality of the results since it is difficult to capture the dynamics of the problem into other mathematical functions. Contig values are computed by applying a final step of refinement with a greedy heuristic regularly used in this domain [10]. We have even found that in some (extreme) cases it is possible that a solution with a better fitness using F than other one generates a larger number of contigs (worse solution). All this suggests that the fitness (overlapping) should be complemented with the actual number of contigs.

$$F(s) = \sum_{i=0}^{N-1} w_{s[i],s[i+1]} \tag{2}$$

However, the calculation of the number of contigs is a quite time-consuming operations, and this definitely precludes any algorithm to use it. A solution to this problem is the utilization of the method which should not need to know the exact number of contigs and thus be computationally light. Our key contribution is to indirectly estimate the number of contigs by measuring the number of contigs that are created or destroyed when tentative solutions are manipulated. We propose a variation of Lin's 2-opt [11] for the DNA field, which does not only use the overlap among the fragments, but that it also takes into account (in an intelligent manner) the number of contigs that have been created or destroyed. The pseudo-code of our proposed method is shown in Algorithm 1.

Algorithm 1. PALS

  $s \leftarrow$ GenerateInitialSolution() {Create the initial solution}
  repeat
    $L \leftarrow \emptyset$
    for i = 0 to N **do**
      for j = 0 to N **do**
        $\Delta_c, \Delta_f \leftarrow$ CalculaeDelta(s,i,j) {See Algorithm 2}
        if $\Delta_c >= 0$ **then**
          $L \leftarrow L \cup < i,j,\Delta_f,\Delta_c >$ {Add candidate movements to L}
        end if
      end for
    end for
    if $L <> \emptyset$ **then**
      $< i,j,\Delta_f,\Delta_c > \leftarrow$ Select(L) {Select a movement among the candidates}
      ApplyMovement(s,i,j) {Modify the solution}
    end if
  until no changes
  return: s

Our algorithm works on a single solution (an integer permutation encoding a sequence of fragment numbers, where consecutive fragments overlap) which is generated using the GenerateInitialSolution method, and it is iteratively modified by the application of movements in a structured manner. A movement is a perturbation (`ApplyMovement` method) that, given a solution s, and two positions i and j, reverses the subpermutation between the positions i and j.

The key step in PALS is the calculation of the variation in the overlap (Δ_f) and in the number of contigs (Δ_c) among the current solution and the resulting solution of applying a movement (see Algorithm 2). This calculation is computationally light since we do not calculate neither the fitness function nor the number of contigs, but instead we estimate the variation of these values. To do this, we only need to analyze the affected fragments by the tentative movement (i, j, $i - 1$ and $j + 1$), removing the overlap score of affected fragments of the current solution and adding the one of the modified solution to Δ_f (equations of lines 4-5 of Algorithm 2) and testing if some current contig is broken (first two if statements of Algorithm 2) or two contigs are merged (last two if statements of Algorithm 2) by the movement operator.

Algorithm 2. `CalculateDelta(s,i,j)` function

$\Delta_c \leftarrow 0$
$\Delta_f \leftarrow 0$
{Calculate the variation in the overlap}
$\Delta_f = w_{s[i-1]s[j]} + w_{s[i]s[j+1]}$ {Add the overlap of the modified solution}
$\Delta_f = \Delta_f - w_{s[i-1],s[i]} - w_{s[j]s[j+1]}$ {Remove the overlap of the current solution}
{Test if a contig is broken, and if so, it increases the number of contigs}
if $w_{s[i-1]s[i]} > cutoff$ **then**
$\quad \Delta_c = \Delta_c + 1$
end if
if $w_{s[j]s[j+1]} > cutoff$ **then**
$\quad \Delta_c = \Delta_c + 1$
end if
{Test if two contig are merged, and if so, it decreases the number of contigs}
if $w_{s[i-1]s[j]} > cutoff$ **then**
$\quad \Delta_c = \Delta_c - 1$
end if
if $w_{s[i]s[j+1]} > cutoff$ **then**
$\quad \Delta_c = \Delta_c - 1$
end if
return: Δ_f, Δ_c

In each iteration, PALS makes these calculations for all possible movements, storing the candidate movements in a list L. Our proposed method only considers candidates to be applied the movements which do not reduce the number of contigs ($\Delta_c \leq 0$). Once it has completed the previous calculations, the method selects a movement of the list L and applies it. The algorithm stops when no more candidate movements are generated.

To complete the definition of our method we must decide how the initial solution is generated (`GenerationInitialSolution` method) and how a movement is selected among all possible candidates (`Select` method). For each one of these operations we propose in this work several versions:

Generation of the Initial Solution:

- **random:** The initial (permutation) solution is randomly generated.
- **greedy:** We begin the permutation with a random fragment and the remaining ones are iteratively assigned maximizing the overlap with respect to the last precedent fragment in the partial permutation.

Selection of the Movements:

- **best:** We select the best movement, i.e., we choose the movement having the lowest Δ_c (thus the movement maintains or reduces the number of contigs). In case that several movements have the same Δ_c, the applied movement will be this with a higher Δ_f value (it increases the overlap among the fragments).
- **first:** This strategy selects the first movement which does not increase the number of contigs ($\Delta_c \leq 0$).
- **random:** This selection method chooses a random movement among all candidate ones.

In the next section, we study the influence of these alternatives on the performance of the method.

4   Experimental Results

In this section we analyze the behavior of our proposed method. First, the target problem instances used are presented in Section 4.1. In the next subsection, we study the influence of the different variations presented in Section 3 in the performance of our algorithm, and finally in Section 4.3, we compare our approach with other assemblers.

The experiments have been executed on a Intel Pentium IV 2.8GHz with 512MB running SuSE Linux 8.1. Because of the stochastic nature of the algorithms, we perform 30 independent runs of each test to gather meaningful experimental data and apply statistical confidence metrics to validate our results and conclusions.

4.1   Target Problem Instances

To test and analyze the performance of our algorithm we generated several problem instances with GenFrag [12]. GenFrag takes a known DNA sequence and uses it as a parent strand from which random fragments are generated according to the criteria supplied by the user (mean fragment length and coverage of parent sequence).

We have chosen four sequences from the NCBI web site[1]: a human MHC class II region DNA with fibronectin type II repeats HUMMHCFIB, with accession number X60189, which is 3,835 bases long; a human apolopoprotein

---

[1] http://www.ncbi.nlm.nih.gov/

HUMAPOBF, with accession number M15421, which is 10,089 bases long; the complete genome of bacteriophage lambda, with accession number J02459, which is 20k bases long; and the Neurospora crassa (common bread mold) BAC, with accession number BX842596, which is 77,292 bases long. The instances generated are free from errors of types 4 and 5 (see Section 2) and the remainder errors are considered and eliminated during the calculation of the overlap among the fragments.

We must remark that the benchmark is large and complex. It is often the case that researches use only one or two instances of low-medium sizes (15-30k bases long). We dare to include two large instances (up to 77k bases long) because the efficiency of our technique, that will be shown to be competitive to modern assemblers.

Table 1. Information of datasets. Accession numbers are used as instance names.

Parameters	Instance								
	X60189				M15421		J02459	BX842596	
Coverage	4	5	5	6	5	7	7	4	7
Mean fragment length	395	386	343	387	398	383	405	708	703
Number of fragments	39	48	66	68	127	177	352	442	773

We experimented with coverage ranging from 4 to 7. The latter instances are very hard since they are generated from very long sequences using a small/medium value of coverage and a very restrictive cutoff (threshold to join adjacent fragments in the same contig). The combination of these parameters produces a very complex instance. For example, longer target sequences have been solved in the literature [5], however they have a higher coverage which makes then not so difficult. The reason is that the coverage measures the redundance of the data, and the higher coverage, the easier the problem. The cutoff, which we have set to thirty (a very high value), provides one filter for spurious overlaps introduced by experimental error. Instances with these features have been only solved adequately when target sequences vary from 20k to 50k base pairs [9,10,13] while we solve instances up to 70k base pairs.

Table 1 presents information about the specific fragments sets we use to test our algorithm.

4.2   Performance Analysis

In this section we analyze the influence of different alternative methods presented in Section 3 on the performance of our method. We study the six versions: the combinations of two solution generation methods (**random** or **greedy**) and three movement selection methods (**best, first** and, **random** movements). We have applied these six methods to solve the eight problem instances presented in Table 1. In Table 2 (accuracy) we include the mean final fitness value and

Table 2. Solution quality (mean fitness and mean number of contigs) for all the instances

Sol. Gen.	random			greedy		
Mov. Sel.	best	first	random	best	first	random
X60189(4)	11451 / 1	11447 / 1	7937 / 3.3	11344 / 1.1	11334 / 1.1	11072 / 1.3
X60189(5)	13932 / 1.5	13897 / 2	11102 / 3.4	13768 / 2.5	13766 / 2.5	13021 / 3.5
X60189(6)	18204 / 1.2	18160 / 1.6	174786 / 3.1	17900 / 1.6	17889 / 1.8	17184 / 2.5
X60189(7)	20968 / 1.5	21051.9 / 1.8	16791 / 3.6	20857 / 2.3	20826 / 2.1	20227 / 3.3
M15421(5)	38454 / 3.6	38370.0 / 4.6	27191 / 10.3	38349 / 5.0	38286 / 5.6	36473 / 9.1
M15421(7)	54666 / 2.8	54852 / 3.2	41182 / 10.3	54344 / 5.1	54393 / 5.3	51609 / 10.7
J02459(7)	115405 / 3.2	115525 / 3.6	81954 / 19.5	114455 / 8.2	114255 / 8.3	109123 / 17.3
BX8425(4)	226744 / 9.9	226363 / 14.1	161891 / 37.6	224656 / 17.9	224689 / 17.5	213589 / 26.4
BX8425(7)	440779 / 7.8	441519 / 10.6	331252 / 42.4	436996 / 22.9	437088 / 22.4	416917 / 37.9

the mean resulting number of contigs, while in Table 3 (efficiency) we show the mean execution time.

Looking at Table 2, the first conclusion is that the method using a random solution generation always achieves a higher accuracy than the one using a greedy generation. The reason of this counterintuitive result is that the greedy generates a high quality solution (especially, for the easiest instances, X60189 and M15421) and then the local search mechanism is not able to further improve it: the method sticks in a local optimum and can not escape from it. In fact, the execution time (Table 3) confirms this hypothesis since we can observe that the execution time using the greedy generation is much lower than the random one, indicating that the former converged very quickly. Analyzing the different movement selection, we can conclude that the selection of the best movement is the most accurate one for all the instances, while the random selection is the worse, producing very low quality solutions (with several tens of contigs for the most complex instances, BX842596). The structured strategy of movements (it performs an ordered improvement in the permutation) followed by the **first** selection seems also to be adequate to this problem, obtaining quite good results only slightly worse than the **best** selection method. Also, we can notice that there are several instances (X60189(4), M15421(7), J02459(7) and, BX842596(7)) where the version which produces the best mean number of contigs is different to the one which obtains the best mean fitness, indicating that the optimization of the overlap among the fragments is not the same as the optimization of the number of contigs (the real objective).

With respect to the execution time, as we stated before, the utilization of the greedy generation allows to reduce the execution time since it converges to suboptimal solutions quickly. With respect to the movement selection method, the random one is the slower since it needs to perform much more movements than the other selection methods. On the other hand, first movement strategy is the fastest since the other methods need to explore all possible movements, while it stops the exploration when it finds a movement which does not increase the number of contigs. Anyway, since all running times are very small, the difference in most of the cases (especially for the easiest ones) are negligible.

Table 3. Mean execution time for all the instances (in seconds)

Solution Gen.	random			greedy		
Movement Sel.	best	first	random	best	first	random
X60189(4)	0.032	0.035	0.034	0.037	**0.031**	0.035
X60189(5)	0.032	0.032	0.042	**0.028**	0.034	0.039
X60189(6)	0.048	**0.039**	0.050	0.043	0.051	0.054
X60189(7)	0.041	0.048	0.054	**0.038**	0.047	0.052
M15421(5)	0.113	0.088	0.167	**0.063**	0.071	0.075
M15421(7)	0.254	0.188	0.389	**0.093**	0.119	0.129
J02459(7)	1.899	1.294	2.494	**0.392**	0.619	0.634
BX842596(4)	3.869	2.475	5.096	1.665	**1.091**	1.119
BX842596(7)	24.82	15.11	28.92	**2.913**	5.955	6.104

4.3   Comparison Against Other Assemblers

Once we have studied our proposed algorithm and how the different alternative methods influence on its performance, we are going to compare its results against other assemblers found in the la literature: a genetic algorithm (GA) [9], a pattern matching algorithm (PMA) [10] and commercially available packages: CAP3 [5] and Phrap [2]. We compare them in terms of the final number of contigs assembled (all these methods use the same cutoff value). Table 4 gives a summary of the results. When a solution with a single contig is achieved, all the algorithms obtain the same solution.

Table 4. Best final number of contig for our assembler (using the best configuration) and for other specialized systems. "-" symbol indicates that this information is not provided by the corresponding paper.

	PALS	GA [9]	PMA [10]	CAP3 [5]	Phrap [2]
X60189(4)	1	1	1	1	1
X60189(5)	1	1	1	1	1
X60189(6)	1	-	1	1	1
X60189(7)	1	1	1	1	1
M15421(5)	1	6	1	2	1
M15421(7)	1	1	2	2	2
J02459(7)	1	13	1	1	1
BX842596(4)	4	-	7	6	6
BX842596(7)	2	-	2	2	2

As it can be seen in Table 4, for X60189 instances (the easiest ones) all the assemblers obtain the optimal number of contigs. However, when the instances are harder, we can notice several differences in the quality of the solutions found by the algorithms. We can conclude that our approach obtains better or similar accuracy than the other methods. In fact, PALS outperforms the remaining tools in two instances. In particular for the BX892596(4), our method represents a new state of the art, and it also achieves the optimum for M15421(7) that was only found by one of the compared systems (Parson's GA [9]).

We do not show execution times because, in general, they are not provided by the authors, but, to give an approximate idea of them to the reader, we can comment that the execution times of these methods range from tens of seconds for the easiest instances to several hours for the hardest ones, while our approach does not spend more than 30 seconds in any instance. However, we should also notice that these tools return the consensus sequence, while PALS returns the ordered layout and an additional step is required to obtain the final consensus string. This final step includes several light operations like the construction of the final DNA sequence from the ordered layout.

5 Conclusions

The DNA fragment assembly is a very complex problem in computational biology. Since the problem is NP-hard, the optimal solution is impossible to find for real cases, except for very small problem instances. Hence, computational techniques of affordable complexity such as heuristics are needed for it.

We have proposed a new problem-aware local search (PALS). Its key contribution is the incorporation of information (an estimation) on the number of contigs into the search mechanism. This feature has allowed us to design a fast and accurate assembler which is competitive against current specialized assemblers. In fact, PALS represents the new state of the art for several complex problem instances. We have also studied the influence of the configuration of our approach. We have observed that the best setting for PALS is to start from a random solution and to select the best movement found in each iteration.

In the future we plan to study our past metaheuristic assemblers augmented with PALS to hopefully solve much larger instances accurately.

Acknowledgments

The authors are partially supported by the Ministry of Science and Technology and FEDER under contract TIN2005-08818-C04-01 (the OPLINK project).

References

1. J. Setubal and J. Meidanis. *Introduction to Computational Molecular Biology*, chapter 4 - Fragment Assembly of DNA, pages 105–139. University of Campinas, Brazil, 1997.
2. P. Green. Phrap. http://www.mbt.washington.edu/phrap.docs/phrap.html.
3. G. G. Sutton, O. White, M. D. Adams, and A. R. Kerlavage. TIGR Assembler: A new tool for assembling large shotgun sequencing projects. *Genome Science & Technology*, pages 9–19, 1995.
4. T. Chen and S. S. Skiena. Trie-based data structures for sequence assembly. In *The Eighth Symposium on Combinatorial Pattern Matching*, pages 206–223, 1998.

5. X. Huang and A. Madan. CAP3: A DNA sequence assembly program. *Genome Research*, 9:868–877, 1999.
6. E. W. Myers. Towards simplifying and accurately formulating fragment assembly. *Journal of Computational Biology*, 2(2):275–290, 2000.
7. P. A. Pevzner. *Computational molecular biology: An algorithmic approach.* The MIT Press, London, 2000.
8. G Luque and E. Alba. Metaheuristics for the DNA Fragment Assembly Problem. *International Journal of Computational Intelligence Research*, 1(2):98–108, January 2006.
9. R. Parsons, S. Forrest, and C. Burks. Genetic algorithms, operators, and DNA fragment assembly. *Machine Learning*, 21:11–33, 1995.
10. L. Li and S. Khuri. A comparison of DNA fragment assembly algorithms. In *International Conference on Mathematics and Engineering Techniques in Medicine and Biological Sciences*, pages 329–335, 2004.
11. S. Lin and B.W. Kernighan. An effective heuristic algorithm for TSP. *Operations Research*, 21:498–516, 1973.
12. M. L. Engle and C. Burks. Artificially generated data sets for testing DNA fragment assembly algorithms. *Genomics*, 16, 1993.
13. Y. Jing and S. Khuri. Exact and heuristic algorithms for the DNA fragment assembly problem. In *Proceedings of the IEEE Computer Society Bioinformatics Conference*, pages 581–2, Stanford Univeristy, August 2003. IEEE Press.

A Hybrid Immune-Based System for the Protein Folding Problem

Carolina P. de Almeida[1], Richard A. Gonçalves[1,2], and Myriam R. Delgado[1]

[1] Federal Technological University of Paraná, Curitiba, PR, Brazil
[2] Department of Computer Science, UNICENTRO, Guarapuava, PR, Brazil

Abstract. This paper describes hybrid algorithms based on artificial immune systems, fuzzy inference systems and tabu search to solve the Protein Folding Problem (PFP) in the 3D Hydrophobic-Polar model, which is a particular instance of the Combinatorial String Folding Problem in a cubic lattice. The proposed methodology aims at enhancing the Clonalg algorithm with a Fuzzy Aging Operator and Weak and Intensive Affinity Maturation. The aging operator uses a fuzzy system to decide which antibodies will be eliminated from the population before the selection stage. The Intensive Maturation employs a Tabu Search strategy. Penalty methods versus feasible search methods are also compared. The proposed hybrid algorithms are tested on a set of standard benchmark instances of PFP and the results attest the efficiency of the methodology.

1 Introduction

In the last few years, the use of hybrid methods inspired by Natural Computing [3] has attracted the attention of many researchers, specially the systems in which two or more methodologies are joined to enhance the final model. In this paper we try to improve the performance of artificial immune systems, known as efficient mechanisms in multi-modal search spaces, by means of local search processes performed during the maturation phase. To maintain the population's diversity and to avoid premature convergence a fuzzy aging operator is also adopted. In this case a fuzzy inference system is used to define the death probability of an antibody. So this natural hybrid system seems to be suitable to solve complex combinatorial problems as the Protein Folding Problem (PFP) considered here.

Proteins are polypeptide chains of amino acid residues. The primary structure of a protein is defined as its linear sequence of amino acids. When left in appropriate environmental conditions, this sequence folds itself, reaching a unique low-energy state. The protein's three-dimensional structure (also called tertiary structure) is determined by this state, which is called the native conformation of the protein. The PFP can be defined as the problem of determining the native conformation of a protein given its primary structure.

Protein folding is a very complex process that involves biological, chemical and physical concepts. Thus, computational methods developed to solve PFPs are generally based on reduced models. Although these reduced models abstract the most relevant features of the whole process, the resulting Protein Folding Problem is still a challenging task. In this paper the Hydrophobic-Polar model in the three-dimensional lattice (3D HP) is adopted.

C. Cotta and J. van Hemert (Eds.): EvoCOP 2007, LNCS 4446, pp. 13–24, 2007.

In this paper we developed different approaches based on Artificial Immune Systems [18], Fuzzy Inference Systems [16] and Tabu Search [10] to solve the Protein Folding Problem. The goal here is to answer the following questions:

- How does the fuzzy aging operator with a weak affinity maturation influence the performance of the Artificial Immune System?
- A penalty-based method could outperform a feasible search approach?
- What is the impact of using the tabu search as an intensive maturation process?

2   The Hydrophobic-Polar Model

Deciding the three-dimensional structure of a protein is fundamental to find out the biological function of such protein. This knowledge may be essential to design new drugs for some kinds of illnesses, to treat or prevent diseases caused by mistakes in the folding process (cystic fibrosis, Alzheimer's, and "mad cow", for example) [6], and to develop biological polymers with specific material properties [17].

Due to the inherited complexity of Protein Folding Problems, simplified models have become very popular. Among several options, the hydrophobic-polar (HP) model [13] is one of the most studied and applied. In the HP model, the twenty amino acids that compose the proteins are divided in two categories: Hydrophobic/Non-polar (H), and Hydrophilic/Polar (P) residues. So, the protein's primary structure can be represented as a string whose elements are in the alphabet $\{H,P\}^+$.

Conformations of an HP sequence are restricted to self avoiding walks on a lattice, since two residues of a protein can not occupy the same position in the lattice space. For the 2D HP model, a two-dimensional square lattice is typically used, while the 3D HP model generally adopts a three-dimensional cubic lattice.

Every feasible conformation in the HP model is associated with a free energy level which is proportional to the number of topological contacts between hydrophobic residu-es that are not neighbors in the given sequence. More specifically, the free energy of a certain conformation with η non-local hydrophobic contacts is $-\eta$.

The HP PFP can be formally defined as follows: given an HP sequence $s = s_1 s_2 \cdots s_n$, we must find an energy-minimizing conformation of s, i.e., find $c^* \in C(s)$ such that $E(c^*) = min\{E(c)|c \in C\}$, where C is the set of all possible conformations and $C(s)$ is the subset of all feasible (self avoiding) conformations for the sequence s [17].

The PFP in this model is known to be NP-hard [1] and is combinatorially equivalent to folding a string of 0's and 1's so that the string forms a self-avoiding walk on the 3D square lattice and the number of adjacent pairs of 1's is maximized [14] [9].

3   Related Works

Most of the researches in PFP are based on the 2D HP model. Monte Carlo methods are considered good algorithms for solving 2D HP PFP. An example of such methodology is the Pruned Enriched Rosenbluth Method (PERM) [11]. Cutello, Nicosia and Pavone applied an Artificial Immune System (AIS) with an aging operator and also achieved good results [8]. Another good algorithm is the Multimeme Algorithm which is capable of solving the 2D HP Protein Folding Problem in different models [12].

More recent works have focused on the 3D HP model. Cotta proposed an Evolutionary Algorithm associated with a Backtracking method [7]. The author compared the results obtained from relative representations with the ones produced by absolute representations. He also compared penalty-based, repair-based, and feasible space approaches. In [17] an Ant Colony Optimization was applied to the 3D HP Protein Folding Problem. The results obtained were favorably compared with state-of-the-art methods. A parallel ACO approach was used in [5]. According to the authors, the parallel approach outperforms single colony implementations both in terms of CPU time and quality of the results. Cutello et al. [9] used an Immune Algorithm based on clonal selection principle with aging operator and memory B cells. The results were compared with those obtained in [7]. In [2] the Tabu Search strategy is applied as the sole method for solving the Protein Folding Problem in the 3D HP model. The results obtained encourages the use of the Tabu Search as a complementary strategy in other approaches.

4   The Proposed Artificial Immune Systems

The algorithms proposed in this paper are based on the Clonalg algorithm [4]. The Clonalg algorithm works with a population of candidate solutions (antibodies), composed of a subset of memory cells (best ones) and a subset of other good individuals. At each generation the n best individuals of the population are selected based on their affinity measures (how good they are as solutions to the problem). The selected individuals are cloned, giving rise to a temporary population of clones. The clones are submitted to an hypermutation operator, whose rate is proportional (or inversely proportional) to the affinity between the antibody and the antigen (the problem to be solved). From this process a maturated antibody population is generated. Some individuals of this temporary population are selected to be memory cells or to be part of the next population. This whole process is repeated until a termination condition is achieved [4]. In the 3D HP Protein Folding Problem the primary representation of the proteins to fold are the antigens and the antibodies are possible conformations in the lattice.

In this work a Fuzzy Aging Operator (which is responsible for eliminating antibodies that are sentenced to death according to the fuzzy inference system), a Weak Affinity Maturation stage (that tries to improve the affinity of the antibodies marked to die by the Fuzzy Aging Operator), and an Intensive Affinity Maturation (that uses the Tabu Search strategy to improve the affinity of the antibodies in the population) are incorporated to the standard Clonalg algorithm. The general form of our algorithms can be summarized by Pseudo-Code 1.

The initial population (generation 0) is randomly generated in two different ways: in the first case only feasible antibodies are possible (i.e., antibodies that represent self avoiding walks of the corresponding sequence in a certain lattice), while the second case accepts infeasible antibodies.

The antibodies are represented using internal coordinates. The internal coordinates depend on the particular lattice topology considered. The representation of the antibodies is better explained in Subsection 4.1.

```
Immune Algorithm (PROT_size, POP_length, dup, MAX_age, HYP_rate)
    generations← 0;
    POP = Initialization();
    Evaluate(POP);
    while (not Terminal_Condition()) do
        POP_c← Cloning(POP,dup);
        POP_h← Hypermutation(POP_c,HYP_rate);
        Evaluate(POP_h);
        POP_m← Hypermacromutation(POP_c);
        Evaluate(POP_m);
        DEATH_POP, POP_a← FuzzyAging(POP,POP_h,POP_m);
        POP_a← WeakMaturation(DEATH_POP);
        POP← Selection(POP_a);
        generations ← generations+1;
        if (Num_Evaliations % Maturation_Evaluations==0)
            POP ← IntensiveMaturation(POP);
    end while
```

Pseudo-Code 1

After being initialized every antibody is evaluated. The affinity of an antibody represents the number of non-local hydrophobic contacts. So, finding the minimal energy of a conformation is transformed into the equivalent problem of maximizing the number of non-local hydrophobic contacts. This is done by the *Evaluate* function, which receives a population of antibodies as a parameter. The evaluation function calculates the number of non-local hydrophobic contacts and the number of collisions.

The *Terminal_Condition* is a function that returns *true* whenever the evolutionary process must be stopped. In this paper the stop criterium is defined as the maximum number of evaluations.

The cloning operator produces some copies (clones) of each antibody. This operator generates an intermediate population of clones (POP_c) with size $POP_{length} * dup$, where POP_{length} is the size of the initial population and dup is the parameter defining the number of copies of each antibody. During the clonal expansion, every cloned antibody inherits the age of its parent.

During the evolutionary process two kinds of hypermutation operators are applied: inversely proportional hypermutation (*Hypermutation*) and *Hypermacromutation*. The *Hypermutation* function receives two parameters - the population of clones and the hypermutation rate (HYP_{rate}) - and returns an intermediate population (POP_h).

In the *Hypermutation* operator, M_{max} (the maximum number of mutations allowed) is inversely proportional to each antibody´s affinity value, and is determined by Eq. 1.

$$M_{max}(A(x)) = \begin{cases} (1 + \frac{E^*}{A(x)}) * \alpha, & \text{if } A(x) > 0 \\ (1 + E^*) * \alpha + \alpha, & \text{if } A(x) = 0 \end{cases} \quad (1)$$

where α = HYP_{rate} * PROT_{size} and $A(x)$ is the affinity value of the individual x and E^* is the best known energy value.

The *Hypermacromutation* function receives just the population of clones and also returns an intermediate population (POP_m). The hypermacromutation tries to mutate each antibody, always generating self avoiding conformations. The maximum number of mutations, that is independent from the affinity of the antibody being hypermacromutated, can be defined as $M_{max} = j - i + 1$, where i and j are two random generated integers such that $(i+1) \leq j \leq \text{PROT}_{size}$. The hypermacromutation operator randomly selects the perturbation direction, either from position i to position j (left to right) or from position j to position i (right to left).

To avoid premature convergence and better explore the search space, we adopted a mechanism to define the actual number of mutations $M \leq M_{max}$ to be applied to an antibody. This mechanism, named First Constructive Mutation (FCM), was first described by Cutello *et. al.* in [9]. The FCM, associated with hypermutation and hypermacromutation operators used here can be described as follows: if the ith mutation in the inversely proportional hypermutation or hypermacromutation gives rise to a feasible individual, the mutation process stops and another process is initiated in the next antibody. Therefore, the effective number of mutations M that occur in an antibody is limited to the range $[1, M_{max}]$, and M is defined as the first mutation that produces a feasible individual. If after M_{max} mutations, no feasible solution is found, the mutated antibody is discarded.

It is important to point out that neither the *Hypermutation* nor the *Hypermacromutation* generates infeasible antibodies (even when the initial population allows infeasible individuals), or permits redundancy in the antibodies receptors. This last restriction imposes that every individual in a population must be different from the others.

An important contribution of this paper regards the analysis of the aging function applied to the antibodies. This function is responsible for determining which antibody (from populations POP, POP_h, and POP_m) should die, must be treated by the Weak Affinity Maturation stage or will be available to take part in the next population. Section 4.2 details the principles of the fuzzy aging operator.

DEATH_POP is formed by antibodies that were sentenced to death while POP_a is formed by the individuals that survived after the aging operator. From this population, the POP_{length} best individuals are chosen (by the *Selection* function) to compose the population of the next generation. An individual (A) is considered better then another (B) if one of three conditions is satisfied: both individuals are feasible and A has a higher number of non-local hydrophobic contacts, or A is feasible and B is not, or both individuals are infeasible and A has a lower number of collisions. If less than POP_{length} individuals survive, new individuals are randomly generated to complete the new population (in the same way as the initial population is generated, i.e., with possible unfeasible individuals in the case of penalty-method).

Finally, at every *Maturation_Evaluations* evaluations the Intensive Affinity Maturation stage based on tabu search is applied to all antibodies of the population. This stage is better described in Section 4.3.

4.1   Representation

As previously discussed, a protein conformation in the HP model is a self avoiding walk of the corresponding sequence in a certain lattice. Then each individual antibody must represent such a walk. This is typically done by using internal coordinates. The internal coordinates depend on the particular lattice topology considered. In this work we used a cubic lattice representation, where each location has at most six neighbors [7].

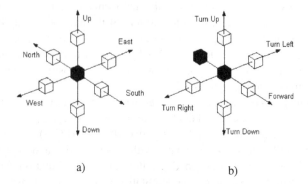

Fig. 1. (a) Absolute moves: the black cube represents the current location; (b) Relative moves: the black cubes represent the current and previous locations

The two major adopted schemes for representing internal movements are (see Figure 1): Absolute and Relative. In the Absolute representation an absolute reference system is assumed and movements are specified accordingly to it. In the cubic lattice a conformation c is represented as a string of size $Prot_{size}$ - 1 over the alphabet {North, South, East, West, Up, Down}. The size of the search space is proportional to the protein's length minus one because the first residue is fixed (this is also adopted by the Relative representation) [19]. The Relative representation has not a fixed reference system: the coordinate system depends on the position of current and previous residues. For the cubic lattice, a conformation c is represented as a string of size $Prot_{size}$ - 1 over the alphabet {Forward, Turn Up, Turn Down, Turn Left, Turn Right} [15].

In this work, the absolute representation was adopted, because previous researches compared both representations and obtained better results with the absolute representation [7] [9].

4.2   Fuzzy Aging Operator

The fuzzy aging operator was developed to avoid premature convergence. This operator contributes to preserve the diversity of the population and to guide the algorithm in the direction of good solutions (as can be seen in Section 5).

The Fuzzy Aging Operator adopted here is inspired by the aging operator proposed by [9] but with a fuzzy inference system used to decide when an antibody should die or be treated. In his work [9], Cutello imposed that every individual had an equal opportunity to explore the search space. This aging operator entirely depended on antibodies's age, regardless of its affinity. Although we think equality is important, we also believe that individuals with higher affinity levels must have better opportunities to remain in the population (an idea that is in accordance with the "survival of the fittest" principle). To accomplish this, we design an aging operator which considers the age, the affinity and the diversity (regarding the remaining individuals of the population) of each antibody. This operator is based on the following assumptions: there must exist good individuals in the population (elitist principle), the diversity must be high (to avoid being

Table 1. Fuzzy Rules used on the Fuzzy Aging Operator

Y and L and L → M	Y and L and L → B	Y and L and L → VB	MA and M and L → L
MA and M and L → M	MA and M and L → B	O and H and L → VL	O and H and L → L
O and H and L → H	Y and L and M → L	Y and L and M → M	Y and L and M → VH
MA and M and M → VL	MA and M and M → L	MA and M and M → H	O and H and M →VL
O and H and M → L	O and H and M → VL	Y and L and H → M	Y and L and H → M
Y and L and H → M	MA and M and H → M	MA and M and H → M	MA and M and H → M
O and H and H → M	O and H and H → M	O and H and H → M	

trapped in local minima and therefore premature convergence), and very old individuals must die (to enrich the diversity of population).

So, a Fuzzy Inference System (FIS) of Mamdani type [16] is being proposed in this paper to define the death probability of an antibody. Such FIS has three input variables: age, affinity and diversity; and one output variable: the death probability. In the adopted knowledge base, linguistic variables have the following set of linguistic terms: T(age) = {Young, Middle Age, Old}; T(affinity) = {Low, Medium, High}; T(diversity) = {Low, Medium, High}; T(death probability) = {Very Low, Low, Medium, High, Very High}. The diversity and affinity variables are normalized in the range [0,1] in a dynamic way, i.e., it is performed at every generation. The FIS output is used to determine if an antibody will survive, must die or is in death eminence and thus must be treated by the Weak Affinity Maturation. Table 1 and Figure 2 illustrate the fuzzy system knowledge base.

4.3 Affinity Maturation Stages

The proposed algorithms have two possible affinity maturation stages: a Weak Affinity Maturation and an Intensive Affinity Maturation. Both stages tries to improve the affinity of an antibody by applying a local search.

Weak Affinity Maturation. This stage is used to improve the affinity of antibodies whose death is imminent.

First of all, Weak Affinity Maturation tries to improve the affinity of an antibody by a scheme analogous to the inversely proportional hypermutation (with the same rate, i.e., both operators has the same M_{max} for each individual), but with some few differences: after the position that is being mutated is defined, the Weak Affinity Maturation tries all the possible new directions (other movements in the lattice space) for this position before proceeding to the next one. Moreover, the process stops only when the affinity of the generated antibody is better than the original one or the maximum number of attempts is achieved. If the first phase of the Weak Affinity Maturation described above is unsuccessful, we also apply another kind of search to this antibody that can be considered a modification of the hypermacromutation operator. At this phase all the possible new directions are tested for every position in the range $[i, j]$ and the process stops only when the affinity of the generated antibody is better than the original one or all positions in the range are tested. After the execution of both phases of the Weak Affinity Maturation, the antibody dies if it is incapable of having its affinity improved.

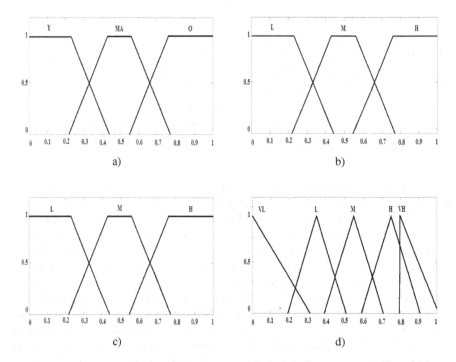

Fig. 2. Membership Functions: a) Age b) Affinity c) Diversity d) Death Possibility

Although the local search mechanism employed by the Weak Affinity Maturation stage is target at improving the quality of the antibodies it´s also responsible for preserving the population´s diversity.

Intensive Affinity Maturation. In order to further improve the quality of the antibodies generated by the proposed AISs, an Intensive Affinity Maturation stage is proposed. This stage performs an intensification process and contrasts with the exploratory nature of the hypermutation and hypermacromutation operators.

In this work, Tabu Search [10] is employed as the local search procedure associated with the Intensive Affinity Maturation and is motivated by the results obtained in [2]. We used just one kind of move: one position mutation (this allows for a smoother progress of the search). The aspiration criteria used is the best ever aspiration criteria, i.e., if a move results in the best individual find so far it´s chosen even if such a move is tabu.

5   Experiments and Results

In this section four variations of the Clonalg [4] algorithm are tested. The features considered here are: Fuzzy Aging Operator, Infeasible Antibodies, Weak Affinity Maturation and Intensive Affinity Maturation.

ClonalgI is the standard Clonalg algorithm while ClonalgII is ClonalgI enhanced with the Fuzzy Aging Operator associated with the Weak Affinity Maturation. ClonalgIII differs from ClonalgII because it allows for the generation of feasible and infeasible antibodies during the initialization and after the aging operator while ClonalgII only permits feasible ones. ClonalgIV enhances ClonalgIII with the use of an Intensive Affinity Maturation process based on Tabu Search.

The algorithms were executed 50 independent times with the following set of parameters (when applicable): 10 individual in the population, duplication rate equals to 4, mutation rate equals to 0.6, tabu tenure equals to 20, 200 Tabu Search Iterations and a termination criterion of 10^5 evaluations. This set of parameters were experimentally determined. It's important to note that the algorithms are not very sensitive to the population size, duplication rate and mutation rate. Similar results are obtained with population sizes varying from 10 to 40, for values greater than 40 and smaller than 10 the quality of the results starts to degenerate. We chose a population size of 10 to accelerate the execution time. Any duplication rate between 2 and 4 are good, but the algorithms run faster with a duplication rate of 4. Mutation rates varying from 0.4 to 0.6 gives similar results, but we chose the value 0.6 due to a slight improvement in the quality of the result without degrading the time performance of the algorithms. Mutation rate values outside this range also degenerates the quality of the results.

To validate our methodology, we compared the proposed algorithms with other evolutionary approaches found in literature [7][9] in the standard benchmark of the 3D HP Protein Folding Problem for 7 different protein sizes (see http://www.cs.sandia.gov/tech_reports/compbio/tortilla-hp-benchmarks.html for more details). The results are summarized on Table 2 in terms of the best found solution (Best), mean, standard deviation (σ) and time (in minutes).

Considering the mean values, all the hybrid systems (ClonalgII, ClonalgIII and ClonalgIV) obtained satisfactory results. In terms of best values, all the proposed approaches found at least equal values when compared with the other evolutionary techniques. In both cases, the best results were obtained by the ClonalgIV. This hybrid system which

Table 2. Results obtained by the proposed algorithms compared to methods of the literature (E^* is the best known values reported on the literature)

	Benchmark Instances							Benchmark Instances						
N.	1	2	3	4	5	6	7	1	2	3	4	5	6	7
Size	20	24	25	36	48	50	60	20	24	25	36	48	50	60
E^*	-11	-13	-9	-18	-29	-26	-49	-11	-13	-9	-18	-29	-26	-49
	Backtracking-EA [7]							Aging-AIS [9]						
Best	-11	-13	-9	-18	-25	-23	-39	-11	-13	-9	-18	-29	-23	-41
Mean	-10.31	-10.90	-7.98	-14.38	-20.80	-20.20	-34.18	-11	-13	-9	-16.76	-25.16	-22.60	-39.28
σ	0	0.36	0	0.88	1.17	1.15	2.00	0	0	0	1.02	0.45	0.40	0.24
T(min)	0.36	0.46	0.5	0.89	2.09	2.34	10.05	0.98	1.59	1.28	2.75	5.83	11.17	13.83
	ClonalgI							ClonalgII						
Best	-11	-13	-9	-18	-29	-27	-48	-11	-13	-9	-18	-29	-30	-47
Mean	-10.40	-11.26	-8.06	-15.04	-24.20	-23.08	-42.65	-11	-12.68	-8.98	-17.20	-26.38	-25.04	-43.04
σ	0.57	0.90	0.87	1.37	2.22	2.05	2.74	0	0.62	0.14	0.90	1.10	1.29	1.29
T(min)	1.43	1.63	1.58	2.24	4.05	4.31	10.34	0.29	0.40	0.43	0.69	2.20	3.96	7.32
	ClonalgIII							ClonalgIV						
Best	-11	-13	-9	-18	-30	-28	-47	-11	-13	-9	-18	**-30**	**-30**	**-51**
Mean	-11	-12.9	9	-17.28	-27.02	-25.06	-44.02	-11	-12.98	-9	*-17.76*	*-28.49*	*-26.36*	*-46.16*
σ	0	0.36	0	0.88	1.17	1.15	2.00	0	0.14	0	0.59	0.92	1.01	1.49
T(min)	0.36	0.46	0.5	0.89	2.09	2.34	10.05	0.98	1.59	1.28	2.75	5.83	11.17	13.83

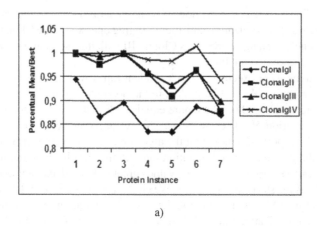

a)

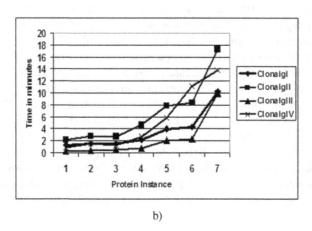

b)

Fig. 3. Graphics of performance: a) Mean/Best(E^*) ratio for each protein b) Mean of the time for one execution of each algorithm to each protein

adopts the fuzzy aging and the intensive maturation process was able to obtain new minimal energy value (E^*_{new} - bold faced in Table 2) for protein instances 5, 6 and 7.

Figure 3 shows the behavior of the mean, calculated as a percentage of the best known energy value described in literature (E^*), and time consuming of the proposed algorithms (both relative to the size of the benchmark instances).

Based on a ranksum test with 95% degree of confiability it is possible to conclude that the quality of the results obtained by ClonalgII are better than those obtained by ClonalgI (Figure 3 a)). But the time spent in the search is almost the double of the time spent by ClonalgI (Figure 3 b)). So it´s reasonable to conclude that the Fuzzy Aging Operator in association with a Weak Affinity Maturation stage is capable of improving the quality of the results by consuming more computational resources. When comparing ClonalgII and III we conclude that although both algorithms achieved similar results in terms of energy (best and mean) values, ClonalgIII is faster (Figure 3 b)). So the use of infeasible individuals in the population is beneficial to the overall performance.

ClonalgIV achieved the best results among all three proposed hybrid algorithms with 95% degree of confiability according to the ranksum test, without requiring excessive computational effort.

ClonalgIV is capable of finding best energy values equal or better than those obtained by a Genetic Algorithm with Backtracking [7] and a Immune Algorithm with Aging Operator [9]. In relation to the mean energy values, ClonalgIV has better values than those obtained in [7] for all instances and is able to produce better results than those presented in [9] for four instances (4, 5, 6 and 7) and worse results only for the second instance.

6 Conclusion

In this paper we proposed three hybrid variations of the Clonalg algorithm. This variations introduced the use of the Fuzzy Aging Operator, the Weak Affinity Maturation and the Intensive Affinity Maturation of the antibodies. The use of infeasible individuals on the population did not degenerate the performance besides it improved the computational time. The Fuzzy Aging Operator - in conjunction with the Weak Affinity Maturation - enhanced the stability of the standard Clonalg algorithm. Finally, the use of the Intensive Affinity Maturation - implemented as a Tabu Search - was able to improve the best and mean energy values and decreased the standard deviation. Clonalg IV - the best hybrid algorithm implemented - allowed us to find energy minima not found by other evolutionary algorithm described in literature.

In future works we intend to analyze the behavior of the proposed algorithms in other combinatorial problems and test the efficiency of other local search strategies as an alternative to the Tabu Search in the Intensive Affinity Maturation Stage.

Acknowledgements

The authors would like to thank the reviewers for the insightful suggestions. Carolina also wants to thank CAPES for financial support.

References

1. B. Berger, and T. Leighton, "Protein Folding in the Hidrophobic-Hidrophilic Model is NP Complete", *Journal of Computational Biology*, v. 5, pp. 27–40, 1998.
2. J. Blazewicz, and P. Lukasiak, and M. Milostan, "Application of tabu search strategy for finding low energy structure of protein," *Artificial Intelligence in Medicine*, v. 35, pp. 135–145, 2005.
3. L. N. de Castro, "Fundamentals of Natural Computing: basic concepts, algorithms, and applications," Chapman & Hall/CRC, 2006.
4. L. N. de Castro, and F. J. Von Zuben, "Learning and Optimization Using the Clonal Selection Principle," *In the Special Issue on Artificial Immune Systems of the journal IEEE Transactions on Evolutionary Computation,* v. 6, n. 3, Jun 2002.
5. D. Chu, M. Till, and A. Y. Zomaya, "Parallel Ant Colony Optimization for 3D Protein Structure Prediction using the HP Lattice Model," *19th International Parallel and Distributed Processing Symposium*, CD-ROM, 2005.

6. F. E. Cohen, and J. W. Kelly, "Therapeutic Approaches to Protein-misfolding Diseases," *Nature*, 426, pp. 905–909, December 2003.

7. C. Cotta, "Protein Structure Prediction Using Evolutionary Algorithms Hybridized with Backtracking," *Proc of the 7th International Work-Conference on Artificial and Natural Neural Networks*, Lecture Notes in Computer Science, 2687, pp. 321–328, 2003.

8. V. Cutello, G. Nicosia, and M. Pavone, "Exploring the Capability of Immune Algorithms: A Characterization of Hypermutation Operators," *Third International Conference on Artificial Immune Systems*, pp. 263–276, Sep. 2004.

9. V. Cutello, G. Morelli, G. Nicosia, and M. Pavone, "Immune Algorithms with Aging Operators for the String Folding Problem and the Protein Folding Problem," *EvoCOP*, pp. 80–90, May. 2005.

10. F. Glover, and M. Laguna , "Tabu Search", in *Modern Heuristic Techniques for Combinatorial Problems*, C. R. Reeves, editor, John Wiley & Sons, Inc, 1993.

11. H. P. Hsu, V. Mehra, W. Nadler, and P. Grassberger, "Growth Algorithm for Lattice Heteropolymers at low Temperatures," *Journal of Chemical Physics*, v. 118, pp. 444–451, 2003.

12. N. Krasnogor, and B. P. Blackburne, and E. K. Burke, and J. D. Hirst "Multimeme Algorithms for Protein Structure Prediction," PPSN VII, Lecture Notes in Computer Science, 2439, pp. 769–778, 2002.

13. K. F. Lau, and K. A. Dill, "Lattice Statistical Mechanics Model of the Conformation and Sequence Space of Proteins," *Macromolecules*, v. 22, pp. 3986–3997, 1989.

14. A. Newman, and M. Ruhl, "Combinatiorial Problems on Strings with Applications to Protein Folding", LATIN'04, pp. 369–378, 2004.

15. A. L. Patton, W. F. Punch III, and E. D. Goodman, "A standard GA approach to native protein conformation prediction," *Proc of 6th International Conference on Genetic Algorithms*, pp. 574–581, 1995.

16. W. Pedricz, and F. Gomide, "An Intruction to Fuzzy Sets: Analysis and Design," Cambridge: MIT Press, 1998.

17. A. Shmygelska, and H. H. Hoos, "An ant colony optimisation algorithm for the 2D and 3D hidrofobic polar protein folding problem," *BMC Bioinformatics*, v. 6, pp. 1–22, Feb. 2005.

18. J. Timmis, and T. Knight, and L. N. de Castro, and E. Hart, "An Overview of Artificial Immune Systems," *In Computation in Cells and Tissues: Perspectives andno Tools for Thought*, pp. 51–86, 2004.

19. R. Unger, and J. Moult, "Genetic algorithms for protein folding simulations," *Journal of Molecular Biology*, v. 231, n. 1, pp. 75–81, 1993.

A Genetic Algorithm for the Resource Renting Problem with Minimum and Maximum Time Lags

Francisco Ballestín

Department of Statistics and OR, Public University of Navarra, Pamplona, Spain
francisco.ballestin@unavarra.es

Abstract. We work with a project scheduling problem subject to temporal constraints where the resource availability costs have to be minimised. As an extension of the more well known Resource Investment Problem, which considers only time-independent costs, this problem includes both time-independent fixed costs and time-dependent variable renting costs for the resources. Consequently, in addition to projects where all resources are bought, we can deal with projects where resources are rented. Based on a new codification of a solution for project scheduling, we develop a Genetic Algorithm capable of outperforming a branch-and-bound procedure that exists for the problem.

Keywords: Project scheduling – Temporal constraints – Resource costs – Metaheuristic algorithms – Genetic Algorithms.

1 Introduction

Resource-constrained project scheduling is concerned with the allocation of time intervals to the processing of activities. The execution of activities requires the use of scarce resources. A classical problem in this field is the resource-constrained project scheduling problem RCPSP (cf. e.g. [3] or [6]) where the objective is to minimise the makespan. The scarcity of resources is given by prescribed limited capacities which must not be exceeded. The RCPSP belongs to the class of problems with regular objective functions. Most of the work in project scheduling has focused on this type of measure of performance. A regular measure of performance is a nondecreasing function of the activity completion times (in the case of a minimization problem). Apart from the minimisation of the makespan, other examples from regular objective functions are the minimization of the mean flowtime, the mean tardiness and the percentage of tardy jobs. In recent years scheduling problems with nonregular measures of performance have gained increasing attention (cf. [16]). A nonregular measure of performance is a measure for which the above definition does not hold. Two popular nonregular measures of performance in the literature are the maximization of the net present value (npv) of the project (cf. [12]) and the minimization of the weighted earliness–tardiness penalty costs of the activities in a project (cf. [16]). In both of these problems, the start times of activities is a key factor in the objective function and the resources are considered, as in the RCPSP, in restrictions. Nevertheless, in some projects the cost of resources is a key factor in itself, even more important than the project length, which should "only" not exceed a certain prefixed limit. One of

C. Cotta and J. van Hemert (Eds.): EvoCOP 2007, LNCS 4446, pp. 25–35, 2007.
© Springer-Verlag Berlin Heidelberg 2007

these problems is the Resource Levelling Problem, RLP. The goal of this problem is to approximately use the same amount of the different types of resource throughout the project, consult e.g. [12] or [2]. Another problem with a resource-based objective function is the resource investment problem RIP, where the use of resources is associated with certain costs which have to be minimised (see e.g. [3] or [13]). Problems of scarce time which have been dealt with in project scheduling literature commonly assume that the costs of making resources available are independent of time. As a consequence, to carry out a project which requires a capacity of x units of a resource, no matter if these x units are used for only one time unit or throughout the whole project execution, the resource availability costs are the same. Hence, making resource units available means buying them. For many real-life projects, however, the use of resources is associated with time-dependent costs, e.g. for heavy machinery or manpower in civil engineering. Moreover, the consideration of time-dependent costs would enable us to model the renting of resources (for resource acquisition via buying and renting see also [1]). That is why the resource renting problem RRP has been proposed (see [14]) where, besides time-independent fixed renting costs, time-dependent variable renting costs are given for the resources.

In that paper it was clear that exact methods are not able to solve medium or large instances. The only heuristic algorithm developed for the RRP is a priority rule in [11]. This paper tries to close this gap by creating a multi-pass algorithm based on that priority rule and especially by developing a metaheuristic algorithm. The developed GA uses a new crossover based on a totally new codification for project scheduling under scarce resources. It also incorporates a local search and a diversification which improve its performance.

In the next section, we introduce the basic terminology of the optimization problem. Section 3 is concerned with priority rules for the problem and a multi-pass algorithm created with them. In Section 4, the developed genetic algorithm is described, together with some extensions that improve its performance. Finally, computational experience with the proposed procedures is presented in Section 5.

2 Preliminaries

2.1 Model of the Problem

In this section we follow [14] and [11]. Let $V = \{0, 1,.., n\}$ be the set of activities of the project, which coincides with the node set of a corresponding activity-on-node project network. The dummy activities 0 and $n+1$ represent the beginning and termination of the project, respectively. Let $p_j \in Z_{\geq 0}$ be the duration (or processing time) and $S_j \in Z_{\geq 0}$ be the start time of activity j where $S_0 = 0$. Then S_{n+1} represents the project duration (or makespan). We assume that there is a prescribed maximum project duration $\overline{d} \in Z_{\geq 0}$, i.e. we have the constraint $S_{n+1} \leq \overline{d}$. If there is a given minimum time lag $d_{ij}^{min} \in Z_{\geq 0}$ between the start of two different activities i and j, i.e., $S_j - S_i \geq d_{ij}^{min}$, we introduce an arc $<i, j>$ in the project network with weight $\delta_{ij} = d_{ij}^{min}$. If there is a given maximum time lag $d_{ij}^{max} \in Z_{\geq 0}$ between the start of activities i and j, i.e., $S_j - S_i \leq d_{ij}^{max}$, we introduce an arc $<j, i>$ with weight $\delta_{ji} = -d_{ij}^{max}$. The arc set of the project network is denoted by E.

The processing of the project activities requires renewable resources. Let R be the set of resources and let $r_{ik} \in Z_{\geq 0}$ ($i \in V$, $k \in R$) be the amount of resource k which is used by activity i in interval $[S_i, S_i + p_i[$. The usage of resources incurs fixed and variable costs. For each unit of resource $k \in R$ rented, we have a fixed renting costs $c_k^f \in Z_{\geq 0}$ arising when bringing the unit into service. In practice, c_k^f often represents a transportation or delivery cost for the resource unit being rented. The variable renting costs of $c_k^v \in Z_{\geq 0}$ refers to one unit of resource k and one unit of time for which the resource unit is rented. Accordingly, the provision of one unit of resource k for a time interval of t time units length leads to fixed costs of c_k^f and to variable costs of $t c_k^v$. We assume that $c_k^f > 0$ or $c_k^v > 0$ for all resources $k \in R$.

Given a schedule S, let $A(S,t) := \{i \in V \mid S_i \leq t < S_i + p_i\}$ be the set of activities in progress at time t and let $r_k(S,t) := \sum_{i \in A(S,t)} r_{ik}$ be the amount of resource k required at time t. Without loss of generality we assume the points in time where the capacities of the resources $k \in R$ can be increased or decreased to be integral. We have to decide on how many units of resource $k \in R$ are to be rented at each point in time $t \in [0, \overline{d}]$. Obviously, at some points in time t it may be optimal to rent more units than used (i.e. more than $r_k(S,t)$ units) in order to reduce the fixed cost. Given schedule S, let $\varphi_k(S,t)$ (or φ_{kt} for short) be the amount of resource k rented at time $t \in [0, \overline{d}]$. Function $\varphi_k(S,.)$ indicates at which points in time resources are allocated or released and thus how long resources are rented. We can restrict ourselves to step functions $\varphi_k(S,.)$ with a finite number of jump discontinuities. Besides, we assume that $\varphi_k(S,.)$ are continuous from the right. Function $\varphi(S,.) := (\varphi_k(S,.))_{k \in R}$ is called a renting policy for a schedule S. Given renting policy $\varphi_k(S,.)$, $c_k^v \int_0^{\overline{d}} \varphi_k(S,t)dt$ represents the total variable renting cost for resource k and planning horizon $\overline{d}$. Let J_k be the finite set of jump discontinuities of function $\varphi_k(S,.)$ on interval $[0, \overline{d}]$ and τ_{min} be the smallest of those jump points. For $t \in J_k \backslash \{\tau_{min}\}$, let $\tau_t := \max\{\tau \in J_k \mid \tau < t\}$ be the largest jump point of function $\varphi_k(S,.)$ less than t and for $t \in J_k$, let

$$\Delta^+ \varphi_{kt} := \begin{cases} [\varphi_k(S,t) - \varphi_k(S,\tau_t)]^+, & \text{if } t > \tau_{min} \\ \varphi_k(S,\tau_{min}), & \text{otherwise} \end{cases} \tag{1}$$

be the increase in the amount or resource k rented at time t. then the total fixed renting cost for resource k equals $c_k^f \sum_{t \in J_k} \Delta^+ \varphi_{kt}$.

Renting policy $\varphi(S,.)$ is called feasible with respect to schedule S if $\varphi_k(S,t) \geq r_k(S,t)$ holds for all $k \in R$ and $t \in [0, \overline{d}]$. Given schedule S, renting policy $\varphi(S,.)$ is called optimal if it is feasible with respect to S and the corresponding total renting cost $\sum_{k \in R} [c_k^v \int_0^{\overline{d}} \varphi_k(S,t)dt + c_k^f \sum_{t \in J_k} \Delta^+ \varphi_{kt}]$ is minimum.

The objective function f of the resource renting problem represents the total renting cost belonging to an optimal renting policy for schedule S and reads as follows

$$f(S) := \sum_{k \in R} \min_{\varphi_k(S,.) \geq r_k(S,.)} \left[c_k^v \int_0^{\bar{d}} \varphi_k(S,t)dt + c_k^f \sum_{t \in J_k} \Delta^+ \varphi_{kt} \right] \qquad (2)$$

Let $\varphi_k^*(S,.)$ be an optimal renting policy for schedule S and $k \in R$. The resource renting problem subject to temporal constraints RRP/max consists of finding a schedule S which satisfies and minimises objective function f. This problem is denoted PS∞ltemp,$\bar{d}$ l$\Sigma\Sigma c_k^v \varphi_{kt} + c_k^f \Delta^+ \varphi_{kt}$ (cf. [11]) and is NP-hard as an extension of the RIP/max (cf. [14]).

It is easy to come up with a renting policy for a feasible schedule S, e.g. $\varphi_k(S,t) = r_k(S,t)$ for all $k \in R$ and $t \in [0, \bar{d}]$. However, it is not straightforward to calculate the optimal renting for S. In [14] it is explained how to create such a renting policy, an algorithm which has $O(n \max(\log n, |R|))$. A similar algorithm to calculate the optimal renting policy is given in [11].

2.2 Candidates for Optimal Solution

In [11], several types of schedules are studied. One of the results states that it is enough to search in the space of quasistable schedules in order to find an optimal solution for the RRP. This means that, when we schedule an activity i in a partial schedule, we only have to look at certain points in time t: (a) we can begin i at the end of a scheduled activity j, (b) we can end i at the beginning of a scheduled activity j, or (c) we can begin i at ES$_i$ or at LS$_i$. ES$_i$ (LS$_i$) denotes the earliest (latest) start time where activity i can be scheduled. Throughout the paper we will only take into account these points at time t when we schedule an activity, but we will not mention it again.

2.3 Test Bed

The tests are based upon a test bed including 3 different sets, UBO10c, UBO20c and UBO50c, with 90 instances with 10, 20 and 50 activities, respectively. The instances have been generated using the problem generator ProGen/max by [10]. The random construction of problem instances by ProGen/max can be controlled by several parameters as the problem size n and $|R|$, the order strength OS of network N as a measure of parallelism, and the resource factor RF as the average fraction of the number of resources used per activity. In addition, the cost quotient CQ denotes the ratio of variable renting costs and fixed renting costs, i.e. $c_k^v = CQ c_k^f$ ($k \in R$). The test set includes 1800 problem instances for each combination of 10, 20 and 50 real activities and 1, 3, and 5 resources. The settings for the order strength, the resource factor, the cost quotient, and the project deadline have been chosen to be $OS \in \{0.25, 0.5, 0.75\}$, $RF = 1$, $CQ \in \{0, 0.1, 0.2, 0.5, 1\}$, and $\bar{d} \in \{d_{0,n+1}, 1.1 d_{0,n+1}, 1.25 d_{0,n+1}, 1.5 d_{0,n+1},\}$, where $d_{0,n+1}$ denotes the length of a longest path from activity 0 to activity $n+1$ in V. Note that the settings for the project deadline ensure that each generated RRP/max-instance possesses a feasible solution. The sets UBO10c and UBO20c were used in [14] and we will compare our best algorithms with the (truncated) B&B from Nübel, B&B$_N$, in these sets. We will use UBO50c in order to compare the different heuristic algorithms we will develop. The quality of an algorithm will be measured by its average deviation with respect to a lower bound calculated by B&B$_N$ at its first iteration. All results refer to a Pentium personal computer with 1.4GHz clock pulse and 512MB RAM.

3 Priority Rules and Multi-pass Algorithms

3.1 Priority Rules

In [11], priority rules are described for PS∞|temp, $\bar{d}$ |f. They build a possible schedule in n steps, at each step an activity is selected and is scheduled locally optimal. The combined priority rule suggested for the RRP is MPA-GRR, where MPA is minimum parallelism first and GRR greatest resource requirements first. The second rule is used as a tie-breaking rule. We have tested this combination against the combination MST-GRR, with MST = minimum slack time ($LS_h - ES_h$). We have also included the random rule (RAN), where an unscheduled random activity is chosen at each iteration. The MPA-GRR rule (44.04% on average) outperforms RAN (46.12% on average), but the best combination is MST-GRR, which obtains 40.15% on average.

3.2 Multi-pass Algorithms

One can create a multi-pass algorithm based on a priority rule by introducing randomness into the procedure. This is done e.g. in the RCPSP (cf. [9]) and the outcome can be used to measure the quality of a more complicated heuristic. We are going to evaluate three methods of introducing randomness: MP1, where only the selection of activities is biased; MP2, where only the schedule of an activity is biased. Each possible point in time to schedule activity i is assigned a probability according to the increase in the objective function obtained if i is scheduled at t; MP3, where both selections are biased. The rule employed in the three algorithms is the one with the best results in the previous section, MST-GRR.

In all the cases we use the regret-based biased random sampling (cf. [4]) and we impose a time limit of 5 seconds. The results say that MP1, the multi-pass algorithm where only the selection of activities is biased, is clearly the best (30.41% on average). MP2 and MP3 obtain an average of 34.06% and 34.73%, respectively.

3.3 Local Search

In the priority rules and MP1, each activity is scheduled locally optimal at each step. However, this optimality is obviously lost when other activities are scheduled. A natural way of improving a schedule obtained by a priority rule is to schedule an activity locally optimal, fixing the rest of the activities. We use this property to create a Local Search and an Improvement Procedure. The first one (LS) unschedules and schedules activities locally optimal until a local optimum is obtained. LS will be used with MP1. The second one (IP) is a faster version where each activity is only chosen and rescheduled once (in a random manner). It will be used with the multi-pass algorithms and the metaheuristic.

In order to test the improvement procedure we have added it to the priority rules and multi-pass algorithm explained in the previous sections. We have observed that the algorithms with IP outperform those without it. Bear in mind that we impose the same time limit on multi-pass algorithms with and without IP, whereas the priority rules with IP need more time than without it (from 0.0058 to 0.0086 seconds if we consider the average of all priority rules without and with IP). It is interesting to note that the three multi-pass algorithms + IP obtain approximately the same results, 29.42%, 29.48% and 29.85% respectively.

4 Genetic Algorithm

In this section we describe the elements of the metaheuristic developed for the RRP/max. Introduced by [7], GAs serve as a heuristic meta strategy to solve hard optimization problems. Following the basic principles of biological evolution, they essentially recombine existing solutions to obtain new ones. The goal is to successively produce better solutions by selecting the better ones of the existing solutions more frequently for recombination. For an introduction into GAs, we refer to [5].

4.1 Codification of a Solution

One of the most important aspects for a genetic algorithm is the codification and decoder used. In our problem we cannot use the usual ones that are employed in many project scheduling problems, for example the activity list and the Serial or Parallel Schedule Generation Scheme in the RCPSP (cf. [9]). The reason that lies behind this is that we do not look for active schedules, a set which always contains an optimal solution for regular objective function (cf. [15]), but for quasistable schedules. We have decided to codify each solution S through a set for each activity $i \in V$, **before**$(i,S) = \{j \in V / S_j + d_j = S_i\}$. That is, before$(i,S)$ is the set of activities that finish exactly when i begins. We also need a set for the schedule S, namely **framework**(S), with framework $(S) = \{i \in V / S_i = d_{0i}$ or $S_i = -d_{i0}\}$, where d_{ij} denotes the length of the longest path between activities i and j if we introduce the arc $<n+1, 0>$ in the original network with weight $\delta_{n+1,0} = - \bar{d}$. To calculate these sets while building a schedule is very straightforward.

4.2 Crossover(M,F)

Another essential part of a genetic algorithm is the crossover operator. Usually, a good crossover operator will be the one capable of transferring (some of) the good qualities from the parents to the children and which can combine them if possible.

In our case, in a schedule with a good objective function there are few points in time where resources have to be newly rented. On the contrary, a bad schedule will have great oscillations in the number of consumed resource units. If a schedule is of good quality it will be then because the order in which activities is scheduled is correct. Note that it is not only necessary that an activity j ends after an activity i, it is important that j ends exactly when i finishes. We have developed a crossover operator that tries to schedule in the children one after the other activities that are scheduled one after the other in the mother M and/or the father F. The pseudo-code for the operator is given below.

Specifically, in order to obtain the daughter D from M and F, we first fix some activities, namely we schedule the activities in framework(M) in the same interval as in

M. Afterwards we perform as many iterations as necessary until all activities have been scheduled, where we select and schedule one activity at each iteration. Sets C and $\overline{C}$ contain, at each point in time, the scheduled and unscheduled activities respectively. Two other sets are essential for the crossover operator, Eleg1 and Eleg2. Both of them must be recalculated at each iteration and are subsets of $\overline{C}$. Eleg1 contains the unscheduled activities i that can be scheduled at that iteration right after a scheduled activity j. However, i must be scheduled right after j in M and in F. Eleg1 also contains the unscheduled activities i that can be scheduled at that iteration right before a scheduled activity j. Activity i must be then scheduled right before j in M and in F. Eleg2 differs from Eleg1 in just one thing: instead of demanding that the activity i is scheduled right after (before) j in M and in F, we simply require this to occur in one of these schedules. It is worthwhile mentioning that Eleg1 do not request the activity j to begin at the same time in M and F.

After defining Eleg1 and Eleg2 we can continue with the description of the crossover operator, which works as follows at each iteration. If Eleg1 is not empty, the procedure randomly selects an activity i from it, and schedules i according to the activity j that has lead to the inclusion of i in Eleg1. That is, if i is right before (after) j in M and F, now it is also scheduled right before (after) j in D. Note that this does not mean that i begins at the same time in D as in both (or any) of the parents. If Eleg1 is empty, we act analogously with Eleg2. We scrutinise the sets in this order so that the daughter inherits structures that are present in both solutions. If both sets Eleg1 and Eleg2 are empty, we randomly choose an unscheduled activity i and schedule it at the best point time. Namely, we look at all possible beginnings for i, calculate the objective function and choose the best alternative. When all activities have been scheduled, we return the solution obtained. The son can be obtained by changing the roles of the mother and the father.

Pseudo-code to obtain the daughter D by recombining the mother M and the father F.

1. $\forall i \in$ framework(M) do $S_i^D = S_i^M$, $\overline{C} = V\backslash$framework(M), $C=\emptyset$.

2. While $\overline{C} \neq \emptyset$

   2.1. Calculate Eleg1. If Eleg1 $\neq \emptyset$, select randomly an activity $i \in$ Eleg1.

   2.2. Else calculate Eleg2. If Eleg2 $\neq \emptyset$, select randomly an activity $i \in$ Eleg2.

   2.3. Else select randomly an activity $i \in \overline{C}$.

   2.4. If $\exists j/\ i \in$ before$(j,M) \cup$ before(j,F), $t^* = S_j^D - p_i$. Else if $\exists j/\ j \in$ before$(i,M) \cup$ before(i,F), $t^*=S_j^D + d_j$. Else choose the best t^* available.

   2.5. Schedule i at t^*, $\overline{C} = \overline{C}\backslash\{i\}$, $C = C \cup \{i\}$. Update ES_i and LS_i $\forall\ i \in \overline{C}$.

3. Return solution D.

Eleg1 = {$i \in \overline{C}$: $\exists\, j \in C$ / $i \in$ before(j,M) $\cap$ before(j,F) with $S_j^D - p_i \in [ES_i, LS_i]$ or $\exists\, j \in C$ / $j \in$ before(i,M) $\cap$ before(i,F) with $S_j^D + p_j \in [ES_i, LS_i]$}.

Eleg2 = {$i \in \overline{C}$: $\exists\, j \in C$ / $i \in$ before(j,M) $\cup$ before(j,F) with $S_j^D - p_i \in [ES_i, LS_i]$ or $\exists\, j \in C$ / $j \in$ before(i,M) $\cup$ before(i,F) with $S_j^D + p_j \in [ES_i, LS_i]$}.

4.3 Crossover(M,F)

The mutation operator also plays an important role in genetic algorithms. Taking advantage of the characteristics of the crossover operator, we have embedded the mutation inside it. Concretely, we calculate a random number in (0, 1) at each iteration. If it is less than a parameter *pmut*, we proceed as if sets Eleg1 and Eleg2 were empty; otherwise the usual steps of the crossover operator are applied. We have fixed *pmut* to 0.1.

4.4 Outline of the Basic Algorithm

We are going to compare two versions of the GA. The outline of the basic GA is the following:

Basic GA

1. *POP* = MP + IP(*nPop*)

2. While the time limit is not reached

    2.1. Divide *POP* randomly in pairs. *POP_New* = $\varnothing$.

    2.2. For each pair (M,F) do:

        2.2.1. Daughter = Crossover(M,F).

        2.2.2. Daughter' = IP(Daughter).

        2.2.3. Son = Crossover(F,M).

        2.2.4. Daughter' = IP(Son).

        2.2.5. POP_New = POP_New $\cup$ {Daughter,Son}.

    2.3. *POP* = Best nPop individuals of *POP* $\cup$ *POP_New*.

3. Return the best solution obtained.

The BasicGA first calculates the initial population with the Multi-Pass algorithm of section 3 plus the improvement procedure described above. Afterwards the same iterations are repeated until the time limit is reached. Firstly, the population is divided into pairs. Secondly, the procedure combines each pair M and F to obtain a daughter and a son with the crossover described in the previous section. Thirdly, the improvement procedure is applied to both solutions. After working with all the pairs we form a new set of solutions with the best *nPop* individuals of the set formed with the old solutions and the new ones.

4.5 Diversification

We have introduced a diversification in the algorithm, based on the function Different(S), which selects $n/2$ activities at random and unschedules them. Afterwards it schedules them randomly one by one. Finally, IP is applied to the new solution. A new population is created through the application of the function Different to each individual of the old population. The new population replaces the old one when certain conditions hold. The conditions are that all the individuals share the same objective function or that *itmax* iterations without improvement of the worst individual in the population have passed. We have fixed *itmax* = 5 in preliminary tests with other instances. We have compared three different versions of the algorithm, all with a time limit of 5 seconds. The diversification improves the genetic algorithm in 0.5%. The best alternative is called GA+D2, which obtains 23.99% on average, has a population size of 12 and calculates 24 solutions in the first step, 12 with MP1+IP and 12 with MP2+IP.

The GA clearly outperforms the multi-pass algorithms with IP. In order to corroborate the quality of the GA, we have added LS to MP1 with a limit of 10 seconds, obtaining an average of 28.35%, more than 1% better than the best MP+IP. However, this percentage is 6% worse than the 22.35% of the basicGA.

5 Comparison with B&B$_N$

In this section we compare the results of priority rule MST, MP1+IP and GA+D2 with the B&B from Nübel (2001), B&B$_N$. Tables 1 and 2 present the results for sets UBO10c and UBO20c, respectively, divided according to CQ. Lines 2-4 of each table show the different average deviations of each algorithm with respect to a lower bound calculated by B&B$_N$ in its first iteration. We have imposed a time limit of 0.5 seconds and 1 second on algorithms GA+D2 and MP1+IP in UBO10c and UBO20c respectively. Algorithm B&B$_N$ has a time limit of 10 seconds on both sets, although the average CPU time is 1.69 and 9.61 seconds respectively. This exact algorithm is not able to find the exact solution in every instance with the given time limit. The last line of Tables 1 and 2 shows the percentage of optimal solutions found by the B&B$_N$. Note that all algorithms have been executed using the same computer.

We can draw several conclusions. There is a completely different situation with 10 and with 20 activities. With 10 activities, the exact algorithm is capable of obtaining the optimal solution in more than 90% of the instances. In this set of instances, this algorithm obtains higher quality solutions in all coefficients than MP1+IP. The priority rules obtain solutions that are much worse than those obtained by B&B$_N$. The metaheuristic and the branch and bound obtain approximately the same solutions' quality for all the coefficients. However, for lower CQ's the latter is slightly better, whereas for the larger CQ's the former is better. With 20 activities, the behavior of B&B$_N$ is much worse. Even the priority rule outperforms it. We can also see that the metaheuristic algorithm is better than the multi-pass algorithm plus IP in all combinations of $\bar{d}$ and CQ.

Table 1. Averages for the different CQ's and total average in UBO10c

CQ	0	0.1	0.2	0.5	1	TOTAL
B&B$_N$	13.79%	25.37%	19.12%	10.25%	5.66%	14.84%
MST	34.83%	34.17%	24.85%	13.05%	7.20%	22.82%
MP1+IP	14.17%	25.87%	19.49%	10.39%	5.71%	15.13%
GA+D2	13.84%	25.42%	19.16%	10.21%	5.62%	14.85%
% opt. sol.	100.00%	93.89%	90.56%	88.33%	86.39%	91.83%

Table 2. Averages for the different CQ's and total average in UBO20c

CQ	0	0.1	0.2	0.5	1	TOTAL
B&B$_N$	64.44%	62.07%	41.67%	20.71%	11.23%	40.02%
MST	55.79%	44.67%	30.66%	15.40%	8.37%	30.98%
MP1+IP	25.26%	35.78%	24.96%	12.57%	6.67%	21.05%
GA+D2	20.58%	33.57%	23.45%	11.70%	6.24%	19.11%
% opt. sol.	14.72%	3.33%	3.06%	2.78%	2.78%	5.33%

6 Summary and Concluding Remarks

In this paper we have developed several multi-pass algorithms and a metaheuristic algorithm for the RRP, a project scheduling problem with a resource-based objective function. This problem enables us to model the renting of resources and is therefore interesting in practice. The metaheuristic algorithm, a GA, relies on a completely new codification of solutions and a crossover operator. Other enhancements of the procedure are a local search method and a diversification. The computational results show that the metaheuristic algorithm is competitive with the existing (truncated) Branch-and-Bound in instances of 10 activities. Besides, the GA already outperforms the truncated B&B when the projects have 20 activities.

Acknowledgements. I am indebted to Hartwig Nübel, for making the exe code of his B&B available through Cristoph Schwindt and Jürgen Zimmermann from the Clausthal University of Technology. I would also like to thank Pilar Lino from the University of Valencia for helping me with the first draft of this paper. Finally, I would like to express my gratitude towards Klaus Neumann from the University of Karlsruhe for suggesting that I studied this problem during my stay at this university. This research was partially supported by the Ministerio de Ciencia y Tecnología under contract TIC2002-02510.

References

1. Ahuja, H.N.: Construction performance control by networks. (1976) John Wiley, New York.
2. Ballestín, F., Schwindt, C., Zimmermann, J.: Resource leveling in make-to-order production: modeling and heuristic solution method, accepted for the International Journal of Operations Research.
3. Demeulemeester, E.: Minimizing resource availability costs in time-limited project networks. Management Science 41 (1995) 1590–1598.
4. Drexl, A.; Scheduling of project networks by job assignment. Management Science, **37** (1991) 1590–1602.
5. Goldberg, D. E.: Genetic algorithms in search, optimization, and machine learning. (1989) Addison-Wesley, Reading, Massachusetts.
6. Herroelen, W., B. De Reyck, and E. Demeulemeester. (1998). "Resource-Constrained Project Scheduling: A Survey of Recent Developments." *Computers and Operations Research* 25(4), 279–302.
7. Holland, H. J.: Adaptation in natural and artificial systems. (1975) University of Michigan Press, Ann Arbor.
8. Kolisch, R. (1995). Project Scheduling under Resource Constraints. Springer.
9. Kolisch, R., Hartmann, S.: Heuristic algorithms for solving the resource-constrained project scheduling problem: Classification and computational analysis, in: J. Weglarz (Ed.), Project Scheduling: Recent Models, Algorithms and Applications, Kluwer Academic Publishers, Berlin, (1999), pp. 147–178.
10. Kolisch, R., Schwindt, C., Sprecher, A.: Benchmark instances for project scheduling problems. In: Weglarz, J. (ed.) Project scheduling – recent models, algorithms and applications, (1999) pp. 197–212. Kluwer, Boston.
11. Neumann, K., Schwindt, C., Zimmermann, J.: Project Scheduling with Time Windows and Scarce Resources. (2003) Springer, Berlin.
12. Neumann, K., Zimmermann, J.: Procedures for resource leveling and net present value problems in project scheduling with general temporal and resource constraints. European Journal of Operational Research, 127 (2000) 425-443.
13. Nübel, H.: Minimierung der Ressourcenkosten für Projekte mit planungsabhängigen Zeitfenstern. (1999) Gabler, Wiesbaden.
14. Nübel, H.: The resource renting problem subject to temporal constraints, OR Spektrum 23 (2001) 359–381.
15. Sprecher, A., Kolisch, R. and Drexl, A.: Semi-active, active and non-delay schedules for the resource-constrained project scheduling problem. European Journal of Operational Research 80 (1995) 94-102.
16. Vanhoucke, M., Demeulemeester, E.L. and Herroelen, W. S.: An exact procedure for the resource-constrained weighted earliness-tardiness project scheduling problem. Annals of Operations Research, 102 (2001) 179–196.

A Probabilistic Beam Search Approach to the Shortest Common Supersequence Problem*

Christian Blum[1], Carlos Cotta[2], Antonio J. Fernández[2], and José E. Gallardo[2]

[1] ALBCOM, Dept. Llenguatges i Sistemes Informàtics
Universitat Politècnica de Catalunya, Barcelona, Spain
cblum@lsi.upc.edu
[2] Dept. Lenguajes y Ciencias de la Computación, ETSI Informática,
Universidad de Málaga, Málaga, Spain
{ccottap,afdez,pepeg}@lcc.uma.es

Abstract. The Shortest Common Supersequence Problem (SCSP) is a well-known hard combinatorial optimization problem that formalizes many real world problems. This paper presents a novel randomized search strategy, called probabilistic beam search (PBS), based on the hybridization between beam search and greedy constructive heuristics. PBS is competitive (and sometimes better than) previous state-of-the-art algorithms for solving the SCSP. The paper describes PBS and provides an experimental analysis (including comparisons with previous approaches) that demonstrate its usefulness.

1 Introduction

The Shortest Common Supersequence Problem (SCSP) is a very well-known problem in the area of string analysis. Basically, the SCSP consists of finding a minimal-length sequence s of symbols from a certain alphabet, such that all strings in a given set L can be *embedded* in s. The resulting combinatorial problem is enormously interesting for several reasons. Firstly, the SCSP constitutes a formalization of different real-world problems. For example, it has many implications in bioinformatics [1]: it is a problem with a close relationship to multiple sequence alignment [2], and to probe synthesis during microarray production [3]. This does not exhaust the practical usefulness of the SCSP though, since it also has applications in planning [4] and data compression [5], among other fields.

Another reason the SCSP has attracted interest lies in its "cleanliness", that is, it is an abstract formulation of different real-world problems that can nevertheless be studied from a theoretical point of view in a context-independent way. Indeed, theoretical computer scientists have analyzed in depth the problem, and we now have accurate characterizations of its computational complexity. These characterizations range from the classical complexity paradigm to the more recent parameterized complexity paradigm. We will survey some of these results in

---

* This work was supported by grants TIN2004-7943-C04-01 and TIN2005-08818-C04-01 of the Spanish government. Christian Blum acknowledges support from the *Ramón y Cajal* program of the Spanish Ministry of Science and Technology.

C. Cotta and J. van Hemert (Eds.): EvoCOP 2007, LNCS 4446, pp. 36–47, 2007.

the next section as well, but it can be anticipated that the SCSP is intrinsically hard [6,7,8] under many formulations and/or restrictions.

The practical impossibility of utilizing exact approaches for tackling this problem in general justifies the use of heuristics. Such heuristic approaches are aimed to producing *probably-* (yet not *provably-*) optimal solutions to the SCSP. Good examples of such heuristics are the MAJORITY MERGE (MM) algorithm, and related variants [9], based on greedy construction strategies. More sophisticated heuristics have been also proposed, for instance, evolutionary algorithms (EAs) [9,10,11,12]. In this work, we present a novel randomized search strategy (or metaheuristic) to tackle the SCSP termed probabilistic beam search (PBS). As the name indicates, this strategy is based in the framework of beam search, but also borrows some heuristic ideas from the greedy constructive heuristics mentioned before. In the following we will show that this strategy can satisfactorily compete in the SCSP arena, outperforming previous state-of-the-art approaches. As a first step, the next section will describe the SCSP in more detail.

2 The Shortest Common Supersequence Problem

First of all, let us introduce some notation that we use in the following. We write $|s|$ for the length of string s ($|s(1)s(2)\ldots s(n)| = n$, where $s(j) \in \Sigma$ is the element at the j-th position of s) and ϵ for the empty string ($|\epsilon| = 0$). Abusing the notation, $|\Sigma|$ denotes the cardinality of set Σ. We use $s \trianglerighteq \alpha$ for the total number of occurrences of symbol α in string s ($s(1)s(2)\ldots s(n) \trianglerighteq \alpha = \sum_{1 \le i \le n, s(i)=\alpha} 1$). We write αs for the string obtained by appending the symbol α in front of string s. Deleting symbol α from the front of string s is denoted by $s|_\alpha$, and is defined as s' when $s = \alpha s'$, or s otherwise. We also use the $|$ symbol to delete a symbol from the front of a set of strings: $\{s_1, \cdots, s_m\}|_\alpha = \{s_1|_\alpha, \cdots, s_m|_\alpha\}$. Finally, $s \in \Sigma^*$ means that s is a finite length string of symbols in Σ.

Let s and r be two strings of symbols taken from an alphabet Σ. String s can be said to be a supersequence of r (denoted as $s \succ r$) using the following recursive definition:

$$\begin{aligned}
s \succ \epsilon &\triangleq \text{True} \\
\epsilon \succ r &\triangleq \text{False}, \quad \text{if } r \ne \epsilon \\
\alpha s \succ \alpha r &\triangleq s \succ r \\
\alpha s \succ \beta r &\triangleq s \succ \beta r, \quad \text{if } \alpha \ne \beta
\end{aligned} \tag{1}$$

Plainly, $s \succ r$ implies that r can be embedded in s, meaning that all symbols in r are present in s in the very same order (although not necessarily consecutive). For example, given the alphabet $\Sigma = \{a, b, c\}$, $aacab \succ acb$. We can now state the SCSP as follows: an instance $I = (\Sigma, L)$ for the SCSP is given by a finite alphabet Σ and a set L of m strings $\{s_1, \cdots, s_m\}$, $s_i \in \Sigma^*$. The problem consists of finding a string s of minimal length that is a supersequence of each string in L ($s \succ s_i, \forall s_i \in L$ and $|s|$ is minimal). For example, given $I = (\{a, b, c\}, \{cba, abba, abc\})$, a shortest common supersequence of I is *abcba*.

The SCSP can be shown to be NP−hard, even if strong constraints are posed on L, or on Σ. For example, it is NP−hard in general when all s_i have length two [5], or when the alphabet size $|\Sigma|$ is two [7]. In principle, these NP−hardness results would have to be approached with caution, since they merely represent a worst case scenario. In this sense, a more sensible characterization of the hardness of the SCSP is provided by the framework of parameterized complexity [13]. This is done by approaching the problem from a multidimensional perspective, realizing its internal structure, and isolating some *parameters*. If hardness (that is, non-polynomial behavior) can be isolated within these parameters, the problem can be *efficiently*[1] solved for fixed values of them. This is the case for several NP−hard problems such as VERTEX COVER [14,15]; the term *fixed-parameter tractable* (FPT) is used to denote these problems. Non-FPT problems will fall under some class in the W−hierarchy. Hardness for class $W[1]$ -the first one above FPT in the hierarchy- is the current measure of intractability: problems in this class cannot be efficiently solved (i.e., in fixed polynomial time) for increasing sizes of the parameter.

Several parameterizations are possible for the SCSP. Firstly, the maximum length k of the supersequence sought can be taken as a parameter. If the alphabet size is constant, or another parameter, then the problem turns in this case to be FPT, since there are at most $|\Sigma|^k$ supersequences, and these can be exhaustively checked. However, this is not very useful in practice because $k \geqslant \max|s_i|$. If the number of strings m is used as a parameter, then SCSP is $W[1]$−hard, and remains so even if $|\Sigma|$ is taken as another parameter [1], or is constant [8]. Failure of finding FPT results in this latter scenario is particularly relevant since the alphabet size in biological problems is fixed (e.g., there are just four nucleotides in DNA). Furthermore, the absence of FPT algorithms implies the absence of fully polynomial-time approximation schemes (FPTAS) for the corresponding problem.

3   Majority Merge Heuristics for the SCSP

The hardness results mentioned previously motivate the utilization of heuristics for tackling the SCSP. One of the most popular algorithms for this purpose is MAJORITY MERGE (MM). This is a greedy algorithm that constructs a supersequence incrementally by adding the symbol most frequently found at the front of the strings in L, and removing these symbols from the corresponding strings. More precisely:

Heuristic MM $(L = \{s_1, \cdots, s_m\})$
1:    let $s \leftarrow \epsilon$
2:    **do**
3:        **for** $\alpha \in \Sigma$ **do let** $\nu(\alpha \mid s) \leftarrow \sum_{s_i \in L, s_i = \alpha s_i'} 1$

---

[1] Here, efficiently means in time $O(f(k)n^c)$, where k is the parameter value, n is the problem size, f is an arbitrary function of k only, and c is a constant independent of k and n.

4:        **let** $\beta \leftarrow \text{argmax}\{\nu(\alpha \mid s) \mid \alpha \in \Sigma\}$

5:        **let** $L \leftarrow L|_\beta$

6:        **let** $s \leftarrow s\beta$

7:    **until** $\sum_{s_i \in L} |s_i| = 0$

8:    **return** s

The myopic functioning of MM makes it incapable of grasping the global structure of strings in L. In particular, MM misses the fact that the strings can have different lengths [9]. This implies that symbols at the front of short strings will have more chances to be removed, since the algorithm has still to scan the longer strings. For this reason, it is less urgent to remove those symbols. In other words, it is better to concentrate in shortening longer strings first. This can be done by assigning a weight to each symbol, depending on the length of the string in whose front is located. Branke *et al.* [9] propose to use precisely this string length as weight, i.e., step 3 in the previous pseudocode would be modified to have

$$\nu(\alpha \mid s) \leftarrow \sum_{s_i \in L, s_i = \alpha s_i'} |s_i'| \tag{2}$$

This modified heuristic is termed WEIGHTED MAJORITY MERGE (WMM), and its empirical evaluation indicates it can outperform MM on some problem instances in which there is no structure, or the structure is deceptive [9,11].

In this work we also consider look-ahead versions of the WMM heuristic. For that purpose we use the notation LA-WMM(l), where $l > 0$ is a parameter that indicates the size (or depth) of the look-ahead. For example, LA-WMM(0) denotes the standard WMM heuristic, whereas LA-WMM(1) is obtained by choosing at each construction step the symbol that corresponds to the first symbol in the best possible sequence of two WMM construction steps. The value of a sequence of two construction steps is obtained by summing the two corresponding WMM weights (see Equation 2). In the following we will refer to these look-head values as the LA-WMM(l) weights.

4 Probabilistic Beam Search for the SCSP

In the following we present a probabilistic beam search (PBS) approach for the SCSP. This algorithm is based on the WMM heuristic outlined before. Beam search is a classical tree search method that was introduced in the context of scheduling [16]. The central idea behind beam search is to allow the extension of partial solutions in more than one way. The version of beam search that we implemented—see algorithm PBS below—works as follows: At each step of the algorithm is given a set B of partial solutions which is called the *beam*. At the start of the algorithm B only contains the empty partial solution ϵ (that is, $B = \{\epsilon\}$). Let C denote the set of all possible children of the partial solutions in B. Note that a child of a string s is obtained by appending one of the symbols from Σ to it. At each step, k_{ext} different (partial) solutions from C are selected; each selection step is either performed probabilistically or deterministically. A chosen (partial) solution is either stored in set B_{compl} in case it is a complete solution, or in the new beam B otherwise. At the end of each construction step the new beam B is reduced in case it contains more than k_{bw} (called the *beam width*) partial solutions. This is done by evaluating the partial solutions in B by means of a lower bound $\text{LB}(\cdot)$, and by subsequently selecting the k_{bw} partial solutions with the smallest lower bound values.

Algorithm PBS($k_{\text{ext}}, k_{\text{bw}}, s_{\text{bsf}}, d$)
1:    **let** $B_{\text{compl}} = \emptyset$
2:    **let** $B = \{\epsilon\}$
3:    **while** $B \neq \emptyset$
4:        **let** $C \leftarrow \text{CHILDREN\_OF}(B)$
5:        **let** $B \leftarrow \emptyset$
6:        **for** $k = 1, \ldots, k_{\text{ext}}$ **do**
7:            **let** $s^t \leftarrow \text{CHOOSE\_FROM}(C, d)$
8:            **if** $\text{LB}(s^t) = |s^t|$ **then**
9:                **let** $B_{\text{compl}} \leftarrow B_{\text{compl}} \cup \{s^t\}$
10:               **if** $|s^t| < |s_{\text{bsf}}|$ **then** $s_{\text{bsf}} \leftarrow s^t$ **endif**
11:           **else**
12:               **if** $\text{LB}(s^t) \leq |s_{\text{bsf}}|$ **then** $B \leftarrow B \cup \{s^t\}$ **endif**
13:           **end if**
14:           **let** $C \leftarrow C \setminus \{s^t\}$
15:       **end for**
16:       **let** $B \leftarrow \text{REDUCE}(B, k_{\text{bw}})$
17:   **end while**
18:   **return** $\text{argmin}\{|s| \,|\, s \in B_{\text{compl}}\}$

In the following we explain the functions of algorithm PBS in more detail. First of all, let us define the following function that will be useful to calculate lower bounds of partial solutions:

$$
\begin{aligned}
s \gg \epsilon &\triangleq (\epsilon, \epsilon) \\
\epsilon \gg r &\triangleq (\epsilon, r), \qquad \text{if } r \neq \epsilon \\
\alpha s \gg \alpha r &\triangleq (\alpha r^e, r^r), \text{ where } (r^e, r^r) = s \gg r \\
\alpha s \gg \beta r &\triangleq s \gg \beta r, \text{ if } \alpha \neq \beta
\end{aligned}
\tag{3}
$$

Intuitively, $s \gg r = (r^e, r^r)$ if r^e is the longest initial segment of string r embedded by s and r^r is the remaining part of r not embedded by s (i.e., $r = r^e r^r$). Note that $s \succ r \iff s \gg r = (r, \epsilon)$.

Function $\text{CHILDREN\_OF}(B)$ produces the set C of all possible children of the partial solutions in B. Note that, given a partial solution s^t, at most $|\Sigma|$ children can be generated by appending each of the symbols from Σ to s^t. Children with unproductive characters (i.e., not contributing to embedding any string in L) are not added to C.

Another important function of algorithm PBS is $\text{CHOOSE\_FROM}(C, d)$. Upon invocation, this function returns one of the partial solutions from set C. This is done as follows. First, we calculate for each $s^t \in C$ a heuristic value $\eta(s^t)$ as follows:

$$\eta(s^t) \leftarrow \left(\sum_{i=1}^{|s^t|} \nu^r \left(s^t(i) \mid s^t(1) s^t(2) \dots s^t(i-1) \right) \right)^{-1}, \tag{4}$$

where $\nu^r(\alpha \mid s)$ is the rank of the weight $\nu(\alpha \mid s)$ which the LA-WMM(l) heuristic assigns to the extension α of string s (see Section 3). The rank of extending string s by symbol α is obtained by sorting all possible extensions of string s with respect to their LA-WMM(l) weights in descending order. Note that the sum shown in Equation 4 is the sum of the ranks of the LA-WMM(l) weights that are used for constructing the partial solution s^t. For example, in case s^t can be constructed by always appending the symbol suggested by the LA-WMM(l) heuristic, the heuristic value of s^t is $\eta(s^t) = \left(\sum_{i=1}^{|s^t|} 1 \right)^{-1} = (|s^t|)^{-1}$. This way of defining the heuristic values has the effect that partial solutions obtained by mostly following the suggestions of the LA-WMM(l) heuristic have a greater heuristic value than others. Given the heuristic values we can define the probability of a (partial) solution s^t from C to be chosen in function CHOOSE_FROM(C, d):

$$\mathbf{p}(s^t) \leftarrow \frac{\eta(s^t)}{\sum_{s^l \in C} \eta(s^l)} \tag{5}$$

However, instead of always choosing a partial solution $s^t \in C$ probabilistically, we employ the following mixed strategy. First, a random number $r \in [0, 1]$ is drawn. If $r < d$ (where $d \in [0, 1]$ is a parameter of the algorithm), the partial solution s^* to be returned by function CHOOSE_FROM(C, d) is selected such that $s^* \leftarrow \text{argmax}\{\mathbf{p}(s^t) \mid s^t \in C\}$. Otherwise, a partial solution is chosen by roulette-wheel-selection using the probabilities defined in Equation 5.[2]

Finally, the lower bound LB(s^t) of a partial solution s^t is calculated as follows: First, we calculate the set of remaining strings in L not embedded by s^t as follows:

$$R(s^t) = \{r_i \mid (s_i^e, r_i) = s^t \gg s_i, s_i \in L\} \tag{6}$$

Let $M(\alpha, R(s^t))$ be the maximum number of occurrences of symbol α in any string in $R(s^t)$:

$$M(\alpha, R(s^t)) = \max\{r_i \unrhd \alpha \mid r_i \in R(s^t)\} \tag{7}$$

Clearly, every common supersequence for the remaining strings must contain at least $M(\alpha, R(s^t))$ copies of the symbol α. Thus a lower bound is obtained by summing the length of the partial solution s^t and the maximum number of occurrences of each symbol of the alphabet in any string in $R(s^t)$:

$$|s^t| + \sum_{\alpha \in \Sigma} M(\alpha, R(s^t)) \tag{8}$$

---

[2] This strategy is known as the *pseudo-random proportional transition rule* in the context of the metaheuristic ant colony optimization.

Note that we use algorithm PBS in a multi-start fashion, that is, given a CPU time limit we apply algorithm PBS over and over again until the CPU limit is reached. The best solution found, denoted by s_{bsf}, is recorded. In fact, this solution is one of the input parameters of algorithm PBS. It is used to exclude partial solutions whose lower bound value exceeds $|s_{bsf}|$ from further consideration.

5   Experimental Evaluation

We implemented our algorithm in ANSI C++ using GCC 3.2.2 for compiling the software. Our experimental results were obtained on a PC with an AMD64X2 4400 processor and 4 Gb of memory.

Two different sets of benchmark instances have been used in the experimentation. The first one—henceforth referred to as SET1—is composed of random strings with different lengths. To be precise, each instance is composed of eight strings, four of them of length 40, and the other four of length 80. Each of these strings is randomly generated, using an alphabet Σ. The benchmark set consists of 5 classes of each 5 instances characterized by different alphabet sizes, namely $|\Sigma| = 2, 4, 8, 16$, and 24. Accordingly, the benchmark set consists of 25 different problem instances. The same instances were used for experimentation, for example, in [11].

A second set of instances is composed of strings with a common source. To be precise, we have considered strings obtained from molecular sequences. The sequences considered comprise both DNA sequences ($|\Sigma| = 4$) and protein sequences ($|\Sigma| = 20$). In the first case, we have taken two DNA sequences of the SARS coronavirus from a genomic database[3]; these sequences are 158 and 1269 nucleotides long. As to the protein sequences, we have considered three of them, extracted from Swiss-Prot[4]:

- *Oxytocin*: quite important in pregnant women, this protein causes contraction of the smooth muscle of the uterus and of the mammary gland. The sequence is 125-aminoacid long.
- *p53*: this protein is involved in the cell cycle, and acts as tumor suppressor in many tumor types; the sequence is 393-aminoacid long.
- *Estrogen*: involved in the regulation of eukaryotic gene expression, this protein affects cellular proliferation and differentiation; the sequence is 595-aminoacid long.

In all cases, problem instances are constructed by generating strings from the target sequence by removing symbols from the latter with probability $p\%$. In our experiments, problem instances comprise 10 strings, and $p \in \{10\%, 15\%, 20\%\}$.

5.1   Algorithm Tuning

First we wanted to find reasonable settings for the parameters of PBS. Remember that PBS has 4 parameters: k_{bw} is the beam width; k_{ext} is the number of

---

[3] http://gel.ym.edu.tw/sars/genomes.html
[4] http://www.expasy.org/sprot/

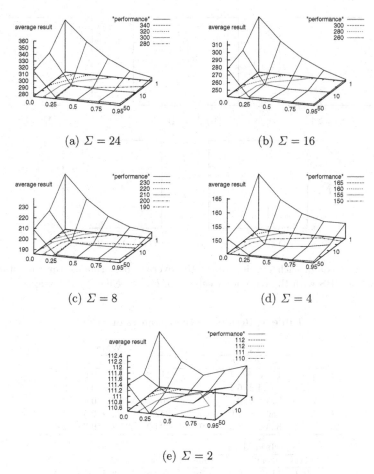

Fig. 1. The z-axis of each graphic shows the average performance of PBS with the parameter settings as specified by the x-axis (parameter d) and the y-axis (parameter k_{bw})

children to be chosen from set C at each step; d is the parameter that controls the extent to which the choice of children from C is performed deterministically. If $d = 1.0$, this choice is always done deterministically, whereas when $d = 0.0$ the choice is always done by roulette-wheel-selection; Finally, l is the depth of the look-ahead function, that is, the parameter in LA-WMM(l) (see Section 3).

In order to reduce the set of parameters to be considered for tuning we decided beforehand to set $k_{ext} = 2 \cdot k_{bw}$. In preliminary experiments we found this setting to be reasonable. Concerning the remaining parameters we tested the following settings: $k_{bw} \in \{1, 10, 50\}$, $d \in \{0.0, 0.25, 0.5, 0.75, 0.95\}$, and $l \in \{0, 1, 2, 3\}$. First we studied the relation between parameters k_{bw} and d, fixing parameter l to the maximum value 3 (that is, $l = 3$). We applied PBS with each combination of parameter values 5 times for 500 CPU seconds to each of the problem instances of

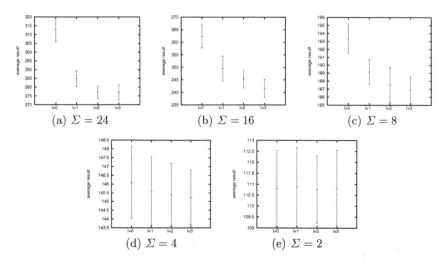

Fig. 2. The y-axis of each graphic shows the average performance (and its standard deviation) of PBS with the parameter setting of l as specified by the x-axis

Table 1. Results for the instances of SET1

$|\Sigma|$	MM			WMM		
	best	mean $\pm \sigma$	i.%	best	mean $\pm \sigma$	i.%
2	112.0	112.0 $\pm$ 0.1	0.0	114.8	114.8 $\pm$ 0.0	-2.5
4	152.6	153.4 $\pm$ 0.7	0.0	157.8	157.8 $\pm$ 0.0	-2.8
8	212.4	213.8 $\pm$ 0.9	0.0	208.2	208.2 $\pm$ 0.0	2.6
16	283.8	286.1 $\pm$ 2.0	0.0	272.8	273.4 $\pm$ 0.5	4.4
24	330.2	333.9 $\pm$ 2.3	0.0	324.0	325.2 $\pm$ 0.7	2.6

$|\Sigma|$	Hybrid MA-BS			PBS		
	best	mean $\pm \sigma$	i.%	best	mean $\pm \sigma$	i.%
2	110.6	110.7 $\pm$ 0.0	1.2	110.8	110.9 $\pm$ 1.7	1.0
4	145.6	146.4 $\pm$ 0.5	4.6	144.8	145.4 $\pm$ 1.5	5.2
8	191.6	192.6 $\pm$ 1.4	9.9	186.4	187.2 $\pm$ 1.7	12.4
16	242.8	244.0 $\pm$ 1.0	14.7	240.4	241.9 $\pm$ 3.4	15.4
24	280.2	281.2 $\pm$ 0.8	15.8	276.4	277.9 $\pm$ 4.0	16.8

SET1. This provided us with 25 results for each instance class (as characterized by the alphabet size). The averaged results for each instance class are shown in the graphics of Figure 1. The results show that, in general, PBS needs some determinism in extension of partial solutions ($d > 0.0$), as well as a beam width greater than 1 ($d > 1$). However, in particular for the problem instances with a smaller alphabet size, the determinism should not be too high and the beam width should not be too big. Therefore, we decided for the settings $d = 0.5$ and $k_{bw} = 10$ for all further experiments.

Finally we performed experiments to decide for the setting of l, that is, the parameter of the look-ahead mechanism. We applied PBS with the four different settings of l ($l \in \{0, 1, 2, 3\}$ 5 times for 500 CPU seconds to each of the problem instances of SET1. This provides us with 25 results for each instance class. The averaged results for each instance class are shown in the graphics of Figure 2. The results show that, in general, the setting of $l = 3$ is best. Especially when the alphabet size is rather large, the performance of PBS is better the higher l is. Only for $\Sigma = 2$, the setting of l does not play much of a role. Therefore, we decided for the setting $l = 3$ for all further experiments.

5.2   Final Experimental Evaluation

We compare the results of PBS to 3 different algorithms: MM refers to a multi-start version of the MM heuristic. This can be done as in case of ties during the solution construction they are broken randomly. Furthermore, WMM refers to a multi-start version of the WMM heuristic, and Hybrid MA-BS refers to

Table 2. Results of the different algorithms for the biological sequences

158-NUCLEOTIDE SARS SEQUENCE

	MM		WMM		Hybrid MA-BS		PBS	
gap%	best	mean ± σ	best	mean ± σ	best	mean ± σ	best	mean ± σ
10%	158	158.0 ± 0.0	158	158.0 ± 0.0	158	158.0 ± 0.0	158	158.0 ± 0.0
15%	160	160.0 ± 0.0	231	231.0 ± 0.0	158	158.0 ± 0.0	158	158.0 ± 0.0
20%	228	229.6 ± 1.8	266	266.0 ± 0.0	158	158.0 ± 0.0	158	158.0 ± 0.0

1269-NUCLEOTIDE SARS SEQUENCE

	MM		WMM		Hybrid MA-BS		PBS	
gap%	best	mean ± σ	best	mean ± σ	best	mean ± σ	best	mean ± σ
10%	1970	2039.9 ± 32.9	2455	2455.0 ± 0.0	1269	1269.0 ± 0.0	1269	1269.0 ± 0.0
15%	2151	2236.4 ± 30.4	2346	2346.0 ± 0.0	1269	1269.0 ± 0.0	1269	1303.8 ± 36.6
20%	2163	2180.2 ± 13.9	2207	2207.0 ± 0.0	1269	1269.0 ± 0.0	1571	1753.2 ± 61.0

125-AMINOACID OXYTOCIN SEQUENCE

	MM		WMM		Hybrid MA-BS		PBS	
gap%	best	mean ± σ	best	mean ± σ	best	mean ± σ	best	mean ± σ
10%	126	126.0 ± 0.0	126	126.0 ± 0.0	125	125.0 ± 0.0	125	125.0 ± 0.0
15%	126	126.0 ± 0.0	126	126.0 ± 0.0	125	125.0 ± 0.0	125	125.0 ± 0.0
20%	132	132.0 ± 0.0	227	227.0 ± 0.0	125	125.0 ± 0.0	125	125.0 ± 0.0

393-AMINOACID P53 SEQUENCE

	MM		WMM		Hybrid MA-BS		PBS	
gap%	best	mean ± σ	best	mean ± σ	best	mean ± σ	best	mean ± σ
10%	393	393.0 ± 0.0	396	396.0 ± 0.0	393	393.0 ± 0.0	393	393.0 ± 0.0
15%	422	422.0 ± 0.0	832	832.0 ± 0.0	393	393.0 ± 0.0	393	393.0 ± 0.0
20%	612	677.1 ± 40.7	833	833.0 ± 0.0	393	393.0 ± 0.0	393	393.0 ± 0.0

595-AMINOACID ESTROGEN SEQUENCE

	MM		WMM		Hybrid MA-BS		PBS	
gap%	best	mean ± σ	best	mean ± σ	best	mean ± σ	best	mean ± σ
10%	628	628.0 ± 0.0	1156	1156.0 ± 0.0	595	595.0 ± 0.0	595	595.0 ± 0.0
15%	671	672.9 ± 2.0	1232	1242.1 ± 4.5	595	595.0 ± 0.0	595	595.0 ± 0.0
20%	1071	1190.3 ± 66.2	1324	1327.9 ± 4.6	595	595.0 ± 0.0	596	596.0 ± 0.0

an algorithm that is a hybrid between beam search and a memetic algorithm. Note that Hybrid MA-BS is a current state-of-the-art technique for the SCSP. The results for all three techniques are taken from [17]. The stopping criterion of MM, WMM, and Hybrid MA-BS was 600 CPU time seconds on a Pentium IV PC with 2400 MHz and 512 MB of memory. This corresponds roughly to the 350 CPU time seconds that we allowed on our machine for PBS.

First, we present the results of PBS for the instances of SET1 in numerical form in Table 1. The results show that PBS is always better than the basic greedy heuristics. With respect to the more sophisticated MA-BS algorithm, the results of PBS are roughly equivalent for $|\Sigma| = 2$. In the remaining instances, PBS improves significantly over the results of Hybrid MA-BS. Even the average performance of PBS is always better than the best performance of Hybrid MA-BS.

As to the biological sequences, the results are shown in Table 2. Again, PBS can be seen to be notoriously better than the greedy algorithms. With respect to MA-BS, PBS is capable of performing at the same level in most instances, systematically finding the optimal solutions. Only in the largest problem instances PBS starts to suffer from the curse of dimensionality. Notice nevertheless that PBS has still room for improvement. For example, using a larger beam width $k_{bw} = 100$ (instead of $k_{bw} = 10$), the results for the two harder SARS DNA instances are notably improved: for 15% gap, the mean result is 1269±0.0 (i.e., systematically finding the optimal solution); for 20% gap, the mean result is 1483±143.1 (best result = 1294) which is much closer to optimal. Further fine-tuning of the parameters may produce even better results.

6   Conclusions and Future Work

We have introduced PBS, a novel metaheuristic that blends ideas from beam search and randomized greedy heuristics. Though relatively simple, and with just four parameters, PBS has been shown to be competitive with a much more complex hybrid metaheuristic for the SCSP that combines beam search and memetic algorithms. Furthermore, PBS has clearly outperformed this latter algorithm in one set of instances. In all cases, PBS has also been shown to be superior to two popular greedy heuristics for the SCSP. In general, PBS is a metaheuristic framework that can be applied to any optimization problem for which exist (1) a constructive mechanism for producing solutions and (2) a lower bound for evaluating partial solutions.

The scalability of PBS is one of the features that deserves further exploration. As indicated by current results, an adequate parameterization of the algorithm can lead to improved results. The underlying greedy heuristic using within PBS, or the probabilistic choosing procedure can be also adjusted. The possibilities are manifold, and work is currently underway in this direction. An additional line of research is the hybridization of PBS with memetic algorithms. A plethora of models are possible in this sense, and using the same algorithmic template of the MA-BA hybrid would be a natural first step.

References

1. Hallet, M.: An integrated complexity analysis of problems from computational biology. PhD thesis, University of Victoria (1996)
2. Sim, J., Park, K.: The consensus string problem for a metric is NP-complete. Journal of Discrete Algorithms **1**(1) (2003) 111–117
3. Rahmann, S.: The shortest common supersequence problem in a microarray production setting. Bioinformatics **19**(Suppl. 2) (2003) ii156–ii161
4. Foulser, D., Li, M., Yang, Q.: Theory and algorithms for plan merging. Artificial Intelligence **57**(2-3) (1992) 143–181
5. Timkovsky, V.: Complexity of common subsequence and supersequence problems and related problems. Cybernetics **25** (1990) 565–580
6. Bodlaender, H., Downey, R., Fellows, M., Wareham, H.: The parameterized complexity of sequence alignment and consensus. Theoretical Computer Science **147** (1–2) (1994) 31–54
7. Middendorf, M.: More on the complexity of common superstring and supersequence problems. Theoretical Computer Science **125** (1994) 205–228
8. Pietrzak, K.: On the parameterized complexity of the fixed alphabet shortest common supersequence and longest common subsequence problems. Journal of Computer and System Sciences **67**(1) (2003) 757–771
9. Branke, J., Middendorf, M., Schneider, F.: Improved heuristics and a genetic algorithm for finding short supersequences. OR-Spektrum **20** (1998) 39–45
10. Branke, J., Middendorf, M.: Searching for shortest common supersequences by means of a heuristic based genetic algorithm. In: Proceedings of the Second Nordic Workshop on Genetic Algorithms and their Applications, Finnish Artificial Intelligence Society (1996) 105–114
11. Cotta, C.: A comparison of evolutionary approaches to the shortest common supersequence problem. In Cabestany, J., Prieto, A., Sandoval, D., eds.: Computational Intelligence and Bioinspired Systems. Volume 3512 of Lecture Notes in Computer Science., Berlin, Springer-Verlag (2005) 50–58
12. Cotta, C.: Memetic algorithms with partial lamarckism for the shortest common supersequence problem. In Mira, J., Álvarez, J., eds.: Artificial Intelligence and Knowledge Engineering Applications: a Bioinspired Approach. Number 3562 in Lecture Notes in Computer Science, Berlin Heidelberg, Springer-Verlag (2005) 84–91
13. Downey, R., Fellows, M.: Parameterized Complexity. Springer-Verlag (1998)
14. Chen, J., Kanj, I., Jia, W.: Vertex cover: further observations and further improvements. In: Proceedings of the 25th International Workshop on Graph-Theoretic Concepts in Computer Science. Number 1665 in Lecture Notes in Computer Science, Berlin Heidelberg, Springer-Verlag (1999) 313–324
15. Niedermeier, R., Rossmanith, P.: A general method to speed up fixed-parameter-tractable algorithms. Information Processing Letters **73** (2000) 125–129
16. Ow, P.S., Morton, T.E.: Filtered beam search in scheduling. International Journal of Production Research **26** (1988) 297–307
17. Gallardo, J.E., Cotta, C., Fernández, A.J.: Hybridization of memetic algorithms with branch-and-bound techniques. IEEE Transactions on Systems, Man, and Cybernetics, Part B (2006) in press.

Genetic Algorithms for Word Problems in Partially Commutative Groups

Matthew J. Craven

Mathematical Sciences, University of Exeter,
North Park Road, Exeter EX4 4QF, UK
m.j.craven@exeter.ac.uk

Abstract. We describe an implementation of a genetic algorithm on
partially commutative groups and apply it to the double coset search
problem on a subclass of groups. This transforms a combinatorial group
theory problem to a problem of combinatorial optimisation. We obtain
a method applicable to a wide range of problems and give results which
indicate good behaviour of the genetic algorithm, hinting at the presence
of a new deterministic solution and a framework for further results.

1 Introduction

1.1 History and Background

Genetic algorithms (hereafter referred to as GAs) were introduced by Holland
[4] and have enjoyed a recent renaissance in many applications including engineering, scheduling and attacking problems such as the travelling salesman and
graph colouring problems. However, the use of GAs in group theory [1,7,8] has
been in operation for a comparatively short time.

This paper discusses an adaptation of GAs for word problems in combinatorial
group theory. We work inside the Vershik groups [11], a subclass of partially
commutative groups (also known as graph groups [10] and trace groups). We
omit a survey of the theory of the groups here and focus on certain applications.

There exists an explicit solution for many problems in this setting. The bi-
automaticity of the partially commutative groups is established in [10], so as a
corollary the conjugacy problem is solvable. Wrathall [12] gave a fast algorithm
for the word problem based upon restricting the problem to a monoid generated
by group generators and their formal inverses. In [13], an algorithm is given for
the conjugacy problem; it is linear time by a stack-based computation model.

Our work is an experimental investigation of GAs in this setting to determine
why they seem to work in certain areas of combinatorial group theory and to determine bounds for what happens for given problems. This is done by translating
given word problems to ones of combinatorial optimisation.

1.2 Partially Commutative Groups and Vershik Groups

Let $X = \{x_1, x_2, \ldots, x_n\}$ be a finite set and define the operation of multiplication
of $x_i, x_j \in X$ to be the juxtaposition $x_i x_j$. As in [13], we specify a *partially*

C. Cotta and J. van Hemert (Eds.): EvoCOP 2007, LNCS 4446, pp. 48–59, 2007.
© Springer-Verlag Berlin Heidelberg 2007

commutative group $G(X)$ by X and the collection of all elements from X that *commute*; that is, the set of all pairs (x_i, x_j) such that $x_i, x_j \in X$ and $x_i x_j = x_j x_i$. For example, take $X = \{x_1, x_2, x_3, x_4\}$ and suppose that $x_1 x_4 = x_4 x_1$ and $x_2 x_3 = x_3 x_2$. Then we denote this group $G(X) = \langle X : [x_1, x_4], [x_2, x_3] \rangle$.

The elements of X are called *generators* for $G(X)$. Note that for general $G(X)$ some generators commute and some do not, and there are no other non-trivial relations between the generators. We concentrate on Vershik groups, a particular subclass of the above groups. For a set X with n elements as above, the *Vershik group of rank n* over X is given by

$$V_n = \langle X : [x_i, x_j] \text{ if } |i - j| \geq 2 \rangle.$$

For example, in the group V_4 the pairs of elements that commute with each other are $(x_1, x_3), (x_1, x_4)$ and (x_2, x_4). We may also write this as $V(X)$ assuming an arbitrary set X. The elements of V_n are represented by group *words* written as products of generators. The *length*, $l(u)$, of a word $u \in V_n$ is the minimal number of single generators from which u can be written. For example $u = x_1 x_2 x_4 \in V_4$ is a word of length three. We use x_i^μ to denote μ successive multiplications of the generator x_i; for example, $x_2^4 = x_2 x_2 x_2 x_2$. Denote the empty word $\varepsilon \in V_n$.

For a subset, Y, of the set X we say the Vershik group $V(Y)$ is a *parabolic subgroup* of $V(X)$. It is easily observed that any partially commutative group G may be realised as a subgroup of a Vershik group V_n of sufficiently large rank n.

Vershik [11] solved the word problem in V_n by means of reducing words to their *normal form*. The Knuth-Bendix normal form of a word $u \in V_n$ of length $l(u)$ may be thought of as the "shortest form" of u and is given by the unique expression

$$\overline{u} = x_{i_1}^{\mu_1} x_{i_2}^{\mu_2} \ldots x_{i_k}^{\mu_k}$$

such that all $\mu_i \neq 0, l(\overline{u}) = \sum |\mu_i|$ and

i) if $i_j = 1$ then $i_{j+1} > 1$;
ii) if $i_j = m < n$ then $i_{j+1} = m - 1$ or $i_{j+1} > m$;
iii) if $i_j = n$ then $i_{j+1} = n - 1$.

The name of the above form follows from the Knuth-Bendix algorithm with ordering $x_1 < x_1^{-1} < x_2 < x_2^{-1} < \ldots < x_n < x_n^{-1}$. We omit further discussion of this here; the interested reader is referred to [6] for a description of the algorithm.

The algorithm to produce the above normal form is essentially a restriction of the stack-based (or heap-based) algorithm of [12] to the Vershik group, and we thus conjecture that the normal form of a word $u \in V_n$ may be computed efficiently in time $O(l(u) \log l(u))$ for the "average case". From now on we write $\overline{u}$ to mean the normal form of the word u. For a word $u \in V_n$, we say that

$$RF(u) = \{x_i^\alpha : l(\overline{ux_i^{-\alpha}}) = l(\overline{u}) - 1, \alpha = \pm 1\}$$

is the *roof of u* and

$$FL(u) = \{x_i^\alpha : l(\overline{x_i^{-\alpha}u}) = l(\overline{u}) - 1, \alpha = \pm 1\}$$

is the *floor of u*. The roof (and floor) of u correspond to the generators which may be cancelled after their inverses are juxtaposed to the right (and left) end of u to create the word u' and u' is reduced to its normal form $\overline{u'}$. For example, if $u = x_1^{-1}x_2x_6x_5^{-1}x_4x_1$ then $RF(u) = \{x_1, x_4\}$ and $FL(u) = \{x_1^{-1}, x_6\}$.

2   Statement of Problem

Given a Vershik group V_n and two words a, b in the group, we wish to determine whether a and b lie in the same double coset with respect to given subgroups. In other words, consider the following problem:

The Double Coset Search Problem (DCSP). Given two parabolic subgroups $V(Y)$ and $V(Z)$ of a Vershik group V_n and two words $a, b \in V_n$ such that $b \in V(Y) \, a \, V(Z)$, find words $x \in V(Y)$ and $y \in V(Z)$ such that $b = xay$.

We attack this group-theoretic problem by transforming it into one of combinatorial optimisation. In the following exposition, an *instance* of the DCSP is specified by a pair (a, b) of given words, each in V_n, and the notation $\mathcal{M}((a,b))$ denotes the set of all *feasible solutions* to the given instance. We will use a GA to iteratively produce "approximations" to solutions to the DCSP, and denote an "approximation" for a solution $(x, y) \in \mathcal{M}((a, b))$ by $(\chi, \zeta) \in V(Y) \times V(Z)$.

Combinatorial Optimisation DCSP

Input: Two words $a, b \in V_n$.
Constraints: $\mathcal{M}((a, b)) = \{(\chi, \zeta) \in \underline{V(Y)} \times V(Z) : \chi a \zeta \doteq b\}$.
Costs: The function $C((\chi, \zeta)) = l(\overline{\chi a \zeta b^{-1}}) \geq 0$.
Goal: Minimise C.

The cost of the pair (χ, ζ) is a non-negative integer imposed by the above function C. The length function defined on V_n takes non-negative values; hence an *optimal solution* for the instance is a pair (χ, ζ) such that $C((\chi, \zeta)) = 0$. Therefore our goal is to minimise the cost function C.

As an application of our work, note that the Vershik groups are inherently related to the braid groups, a rich source of primitives for algebraic cryptography. In particular, the DCSP in the Vershik groups is an analogue of an established braid group primitive. The reader is invited to consult [5] for further details.

In the next section we expand these notions and detail the method we use to solve this optimisation problem.

3   Genetic Algorithms on Vershik Groups

3.1   An Introduction to the Approach

For brevity we do not discuss the elementary concepts of GAs here, but refer the reader to [4,9] for a discussion of GAs and remark that we use standard terms such as *cost-proportionate selection* and *reproductive method* in a similar way.

We give a brief introduction to our approach. We begin with an initial population of "randomly generated" pairs of words, each pair of which is treated as an
approximation to a solution $(x, y) \in \mathcal{M}((a, b))$ of an instance (a, b) of the DCSP.
We explicitly note that the GA does not know either of the words x or y. Each
pair of words in the population is ranked according to some cost function which
measures how "closely" the given pair of words approximates (x, y). After that
we systematically imitate natural selection and breeding methods to produce a
new population, consisting of modified pairs of words from our initial population.
Each pair of words in this new population is then ranked as before. We continue
to iterate populations in this way to gather steadily closer approximations to a
solution (x, y) until we arrive at a solution (or otherwise).

3.2   The Representation and Computation of Words

We work in V_n and two given parabolic subgroups $V(Y)$ and $V(Z)$, and wish
the GA to find an exact solution to a posed problem. We naturally represent a
group word $u = x_{i_1}^{\mu_1} x_{i_2}^{\mu_2} \ldots x_{i_k}^{\mu_k}$ of arbitrary length by a string of integers, where
we consecutively map each generator of the word u as follows:

$$x_i^{\epsilon_i} \rightarrow \begin{cases} +i & \text{if } \epsilon_i = +1 \\ -i & \text{if } \epsilon_i = -1 \end{cases}$$

For example, if $u = x_1^{-1} x_4 x_2 x_3^2 x_7 \in V_7$ then u is represented by the string
-1 4 2 3 3 7. In this context the length of u is equal to the number of integers
in its string representation. We define a *chromosome* to be the GA representation
of a pair (χ, ζ) of words, and note that each word is naturally of variable length.
Moreover a *population* is a multiset of a fixed number p of chromosomes. The GA
has two populations in memory, the *current population* and the *next generation*.
As with traditional GAs the current population contains the chromosomes under
consideration at the current iteration of the GA, and the next generation has
chromosomes deposited into it by the GA which form the current population on
the next iteration. A *subpopulation* is a submultiset of a given population.

We use the natural representation for ease of algebraic operation, acknowl-
edging that faster or more sophisticated data structures exist, for example the
stack-based data structure of [13]. However we believe the simplicity of our representation yields relatively uncomplicated reproductive algorithms. In contrast,
we believe a stack-based data structure yields reproductive methods of considerable complexity. We give our reproductive methods in the next subsection.

Besides normal form reduction of a word u we use *pseudo-reduction* of u. Let
$\{ x_{i_{j_1}}, x_{i_{j_1}}^{-1}, \ldots, x_{i_{j_m}}, x_{i_{j_m}}^{-1} \}$ be the generators which would be removed from u if
we were to reduce u to normal form. Pseudo-reduction of u is defined as simply
removing the above generators from u. There is no reordering of the resulting
word (as with normal form). For example, if $u = x_6 x_8 x_1^{-1} x_2 x_8^{-1} x_2^{-1} x_6 x_4 x_5$ then
its *pseudo-normal form* is $\tilde{u} = x_6 x_1^{-1} x_6 x_4 x_5$ and the normal form of u is $\bar{u} = x_1^{-1} x_4 x_6^2 x_5$. Clearly, we have $l(\tilde{u}) = l(\bar{u})$. This form is efficiently computable,
with complexity at most that of the algorithm used to compute the normal form
$\bar{u}$. Note, a word is not assumed to be in any given form unless otherwise stated.

3.3   Reproduction

The following reproduction methods are adaptations of standard GA reproduction methods. The methods act on a subpopulation to give a child chromosome, which we insert into the next population (more details are given in section 5).

1. Sexual (*crossover*): by some selection function, input two parent chromosomes c_1 and c_2 from the current population. Choose one random segment from c_1, one from c_2 and output the concatenation of the segments.
2. Asexual: input a parent chromosome c, given by a selection function, from the current population. Output one child chromosome by one of the following:
   (a) *Insertion* of a random generator into a random position of c.
   (b) *Deletion* of a generator at a random position of c.
   (c) *Substitution* of a generator located at a random position in c with a random generator.
3. Continuance: return several chromosomes $c_1, c_2, \ldots, c_m$ chosen by some selection algorithm, such that the first one returned is the "fittest" chromosome (see the next subsection). This method is known as *partially elitist*.
4. Non-Local Admission: return a random chromosome by some algorithm.

With the exception of continuance, the methods are repeated for each child chromosome required.

3.4   The Cost Function

In a sense, a cost function induces a partial metric over the search space to give a measure of the "distance" of a chromosome from a solution. Denote the solution of an instance of the DCSP in section 2 by (x, y) and a chromosome by (χ, ζ). Let $E(\chi, \zeta) = \chi a \zeta b^{-1}$; for simplicity we denote this expression by E. The normal form of the above expression is denoted $\overline{E}$. When (χ, ζ) is a solution to an instance, we have $\overline{E} = \varepsilon$ (the empty word) with defined length $l(\overline{E}) = 0$.

The cost function we use is as follows: given a chromosome (χ, ζ) its cost is given by the formula $C((\chi, \zeta)) = l(\overline{E})$. This value is computed for every chromosome in the current population at each iteration of the GA. This means we seek to minimise the value of $C((\chi, \zeta))$ as we iterate the GA.

3.5   Selection Algorithms

We realise continuance by roulette wheel selection. This is cost proportionate. As we will see in Algorithm 2, we implicitly require the population to be ordered best cost first. To this end, write the population as a list $\{(\chi_1, \zeta_1), \ldots, (\chi_p, \zeta_p)\}$ where $C(\chi_1, \zeta_1) \leq C(\chi_2, \zeta_2) \leq \ldots \leq C(\chi_p, \zeta_p)$. Then the algorithm is as follows:

Algorithm 1 (Roulette Wheel Selection)

> INPUT: *The population size p; the population chromosomes (χ_i, ζ_i); their costs $C((\chi_i, \zeta_i))$; and n_s, the number of chromosomes to select*

OUTPUT: n_s *chromosomes from the population*

1. *Let* $W \leftarrow \sum_{i=1}^{p} C((\chi_i, \zeta_i))$;
2. *Compute the sequence* $\{p_s\}$ *such that* $p_s((\chi_i, \zeta_i)) \leftarrow \frac{C((\chi_i, \zeta_i))}{W}$;
3. *Reverse the sequence* $\{p_s\}$;
4. *For* $j = 1, \ldots, p$, *compute* $q_j \leftarrow \sum_{i=1}^{j} p_s((\chi_i, \zeta_i))$;
5. *For* $t = 1, \ldots, n_s$, *do*
   (a) *If* $t = 1$ *output* (χ_1, ζ_1), *the chromosome with least cost. End.*
   (b) *Else*
       i. *Choose a random* $r \in [0, 1]$;
       ii. *Output* (χ_k, ζ_k) *such that* $q_{k-1} < r < q_k$. *End.*

The algorithm respects the requirement that chromosomes with least cost are selected more often. For crossover we use *tournament selection*, where we input three randomly chosen chromosomes in the current population and select the two with least cost. If all three have identical cost, then select the first two chosen. Selection of chromosomes for asexual reproduction is at random from the current population.

4 Traceback

In many ways, cost functions are a large part of a GA. But the reproduction methods often specify that a random generator is chosen, so reducing the number of possible choices of generator may serve to guide the GA. We give a possible approach to reducing this number and term it *traceback*. In brief, we take the problem instance given by the pair (a, b) and use a and b to determine properties of a feasible solution $(x, y) \in \mathcal{M}((a, b))$ to the instance. This approach exploits the "geometry" of the search space by tracking the process of reduction of E to its normal form in V_n and proceeds as follows:

Recall Y and Z respectively denote the set of generators of the parabolic subgroups $G(Y)$ and $G(Z)$. Suppose we have a chromosome (χ, ζ) at some stage of the GA computation. Form the expression $E = \chi a \zeta b^{-1}$ associated to the given instance of the DCSP and label each generator from χ and ζ with its position in the product $\chi\zeta$. Then reduce E to its normal form $\overline{E}$; during reduction the labels travel with their associated generators. As a result some generators from χ or ζ may be cancelled or not, and the set of labels of the non-cancelled generators of χ and ζ give the original positions.

The generators in V_n which commute mean that the chromosome may be split into *blocks* $\{\beta_i\}$. Each block is formed from at least one consecutive generator of χ and ζ which move together under reduction of E. Let B be the set of all blocks from the above process. Now a block $\beta_m \in B$ and a position q (which we call the *recommended position*) at either the left or right end of that block are randomly chosen. Depending upon the position chosen, take the subword δ between either the current and next block β_{m+1} or the current and prior block β_{m-1} (if available). If there is just one block, then take δ to be between β_1 and the end or beginning of $\overline{E}$.

Then identify the word χ or ζ from which the position q originated and its associated generating set $S = Y$ or $S = Z$. The position q is at either the left or right end of the chosen block. So depending on the end of the block chosen, randomly select the inverse of a generator from $RF(\delta) \cap S$ or $FL(\delta) \cap S$. Call this the *recommended generator* g. Note if both χ and ζ are entirely cancelled (and so B is empty), we return a random recommended generator and position.

With these, the insertion algorithm inserts the inverse of the generator on the appropriate side of the recommended position in χ or ζ. In the cases of substitution and deletion, we substitute the recommended generator or delete the generator at the recommended position. We now give an example for the DCSP on V_{10} with the two parabolic subgroups of $V(Y) = V_7$ and $V(Z) = V_{10}$.

Example of Traceback on a Given Instance. Take the short DCSP instance

$$(a, b) = (x_2^2 x_3 x_4 x_5 x_4^{-1} x_7 x_6^{-1} x_9 x_{10}, \; x_2^2 x_4 x_5 x_4^{-1} x_3 x_7 x_6^{-1} x_{10} x_9)$$

and let the current chromosome be $(\chi, \zeta) = (x_3 x_2^{-1} x_3^{-1} x_5 x_7, \; x_5 x_2 x_3 x_7^{-1} x_{10})$. Represent the labels of the positions of the generators in χ and ζ by the following numbers immediately above each generator:

$$\begin{array}{ccccc|ccccc}
0 & 1 & 2 & 3 & 4 & 5 & 6 & 7 & 8 & 9 \\
x_3 & x_2^{-1} & x_3^{-1} & x_5 & x_7 & x_5 & x_2 & x_3 & x_7^{-1} & x_{10}
\end{array}$$

Forming E and reducing it to its Knuth-Bendix normal form gives

$$\overline{E} = \begin{array}{ccccccccccc}
0 & 1 & 2 & & & 3 & & & & 4 & \\
x_3 & x_2^{-1} & x_3^{-1} & x_2 & x_2 & x_3 & x_2^{-1} & x_5 & x_4 & x_5 & x_4^{-1} & x_7 & x_7 \\
& 5 & & & 8 & & & & 9 & & \\
x_6^{-1} & x_5 & x_4 & x_7^{-1} & x_6 & x_5^{-1} & x_4^{-1} & x_7^{-1} & x_9 & x_{10} & x_{10} & x_9^{-1} & x_{10}^{-1}
\end{array}$$

which contains eight remaining generators from (χ, ζ). Take cost to be $C((\chi, \zeta)) = l(\overline{E}) = 26$, the number of generators in $\overline{E}$ above. There are three blocks for χ:

$$\beta_1 = \begin{array}{ccc} 0 & 1 & 2 \\ x_3 & x_2^{-1} & x_3 \end{array}, \; \beta_2 = \begin{array}{c} 3 \\ x_5 \end{array}, \; \beta_3 = \begin{array}{c} 4 \\ x_7 \end{array}$$

and three for ζ:

$$\beta_4 = \begin{array}{c} 5 \\ x_5 \end{array}, \; \beta_5 = \begin{array}{c} 8 \\ x_7^{-1} \end{array}, \; \beta_6 = \begin{array}{c} 9 \\ x_{10} \end{array}$$

Suppose we choose position eight, which is in ζ and is block β_5. This is a block of length one; we may take the word to the left or the right as our choice for δ.

Suppose we choose the word to the right, so $\delta = x_6 x_5^{-1} x_4^{-1} x_7^{-1} x_9 x_{10}$ and in this case, $S = \{x_1, \ldots, x_{10}\}$. So we choose a random generator from $FL(\delta) \cap S = \{x_6, x_9\}$. Choose $g = x_6^{-1}$ and so ζ becomes $\zeta' = x_5 x_2 x_3 x_7^{-1} x_6^{-1} x_{10}$, with $\chi' = \chi$. The cost becomes $C((\chi', \zeta')) = l(\overline{\chi' a \zeta' b^{-1}}) = 25$. Note that we could have taken any block and the permitted directions to create δ. In this case, there are eleven choices of δ, clearly considerably fewer than the total number of subwords of $\overline{E}$. Traceback provides a significant increase in performance over merely random selection (this is easily calculated in the above example to be by a factor of 38).

5 Setup of the Genetic Algorithm

5.1 Specification of Output Alphabet

Let $n = 2m$ for some integer $m > 1$. Define the subsets of generators $Y = \{x_1, \ldots, x_{m-1}\}$, $Z = \{x_{m+2}, \ldots, x_n\}$ and two corresponding parabolic subgroups $G(Y) = \langle Y \rangle, G(Z) = \langle Z \rangle$. Clearly $G(Y)$ and $G(Z)$ commute as groups: if we take any $m > 1$ and any words $x_y \in G(Y)$, $x_z \in G(Z)$ then $x_y x_z = x_z x_y$. We direct the interested reader to [5] for information on the importance of the preceding statement. Given an instance (a, b) of the DCSP with parabolic subgroups as above, we will seek a representative for each of the two words $x \in G(Y)$ and $y \in G(Z)$ that are a solution to the DCSP. Let us label this problem (P).

5.2 The Algorithm and Its Parameters

Given a chromosome (χ, ζ) we choose crossover to act on either χ or ζ at random, and fix the other component of the chromosome. Insertion is performed according to the position in χ or ζ given by traceback and substitution is with a random generator, both such that if the generator chosen cancels with a neighbouring generator from the word then another random generator is chosen. We choose to use pseudo-normal form for all chromosomes to remove all redundant generators while preserving the internal ordering of (χ, ζ).

By experiment, GA behaviour and performance is mostly controlled by the *parameter set* chosen. A parameter set is specified by the population size p and numbers of children begat by each reproduction algorithm. The collection of numbers of children is given by a multiset of non-negative integers $P = \{p_i\}$, where $\sum p_i = p$ and each p_i is given, in order, by the number of crossovers, substitutions, deletions, insertions, selections and random chromosomes. The GA is summarised by the following algorithm:

Algorithm 2 (GA for DCSP)

INPUT: *The parameter set, words a, b and their lengths $l(a), l(b)$, suicide control σ, initial length L_I*

OUTPUT: *A solution (χ, ζ) or timeout; i, the number of populations*

1. *Generate the initial population P_0, consisting of p random (unreduced) chromosomes (χ, ζ) of initial length L_I;*
2. *$i \leftarrow 0$;*
3. *Reduce every chromosome in the population to its pseudo-normal form.*
4. *While $i < \sigma$ do*

    (a) *For $j = 1, \ldots, p$ do*

        i. *Reduce each pair $(\chi_j, \zeta_j) \in P_i$ to its pseudo-normal form $(\tilde{\chi}_j, \tilde{\zeta}_j)$;*

        ii. *Form the expression $E = \tilde{\chi}_j \, a \, \tilde{\zeta}_j \, b^{-1}$;*

        iii. *Perform the traceback algorithm to give $C((\chi_j, \zeta_j))$, recommended generator g and recommended position q;*

> (b) *Sort current population P_i into least-cost-first order and label the chromosomes $(\tilde{\chi}_1, \tilde{\zeta}_1), \ldots, (\tilde{\chi}_p, \tilde{\zeta}_p)$;*
> (c) *If the cost of $(\tilde{\chi}_1, \tilde{\zeta}_1)$ is zero then return solution (χ_1, ζ_1) and END.*
> (d) *$P_{i+1} \leftarrow \emptyset$;*
> (e) *For $j = 1, \ldots, p$ do*
>> i. *Using the data obtained in step 4(a)(iii), perform the appropriate reproductive algorithm on $(\tilde{\chi}_j, \tilde{\zeta}_j)$ and denote the result (χ'_j, ζ'_j);*
>> ii. *$P_{i+1} \leftarrow P_{i+1} \cup \{(\chi'_j, \zeta'_j)\}$;*
> (f) *$i \leftarrow i + 1$.*
> 5. *Return failure.* END.

The positive integer σ is an example of a *suicide control*, where the GA stops (suicide) if more than σ populations have been generated. In all cases here, σ is chosen by experimentation; GA runs that continued beyond σ populations were unlikely to produce a successful conclusion. By deterministic search we found a population size of $p = 200$ and parameter set $P = \{5, 33, 4, 128, 30, 0\}$ for which the GA performs well when $n = 10$. We observed that the GA exhibits the well-known common characteristic of sensitivity to changes in parameter set; we consider this in future work. We found an optimal length of one for each word in our initial population, and now devote the remainder of the paper to our results of testing the GA and analysis of the data collected.

5.3   Method of Testing

We wished to test the performance of the GA on "randomly generated" instances of problem (P). Define the length of an instance of (P) to be the set of lengths $\{l(\bar{a}), l(\bar{x}), l(\bar{y})\}$ of words $a, x, y \in V_n$ used to create that instance. Each of the words a, x and y are generated by simple random walk on V_n. To generate a word $\bar{u}$ of given length $k = l(\bar{u})$ firstly generate the unreduced word u_1 with unreduced length $l(u_1) = k$. Then if $l(\overline{u_1}) < k$, generate u_2 of unreduced length $k - l(\overline{u_1})$, take $u_1 u_2$ and repeat this procedure until we produce a word $u = u_1 u_2 \ldots u_r$ with $l(\bar{u})$ equal to the required length k.

We identified two key input data for the GA: the length of an instance of (P) and the group rank, n. Two types of tests were performed, varying these data:

1. Test of the GA with long instances while keeping the rank small;
2. Test of the GA with instances of moderate length while increasing the rank.

The algorithms and tests were developed and conducted in GNU C++ on a Pentium IV 2.53 GHz computer with 1GB of RAM running Debian Linux 3.0.

5.4   Results

Define the *generation count* to be the number of populations (and so iterations) required to solve a given instance; see the counter i in Algorithm 2. We present the results of the tests and follow this in section 5.5 with discussion of the results.

Table 1. Results of increasing instance lengths for constant rank $n = 10$

Instance	$l(\overline{a})$	$l(\overline{x})$	$l(\overline{y})$	$\overline{g}$	$\overline{t}$	σ_g	sec/gen
I1	128	16	16	183	59	68.3	0.323
I2	128	32	32	313	105	198.5	0.339
I3	256	64	64	780	380	325.5	0.515
I4	512	64	64	623	376	205.8	0.607
I5	512	128	128	731	562	84.4	0.769
I6	1024	128	128	1342	801	307.1	0.598
I7	1024	256	256	5947	5921	1525.3	1.004
I8	2048	512	512	14805	58444	3576.4	3.849

Increasing Length. We tested the GA on eight randomly generated instances (I1)–(I8) with the rank of V_n set at $n = 10$. The instances (I1)–(I8) were generated beginning with $l(\overline{a}) = 128$ and $l(\overline{x}) = l(\overline{y}) = 16$ for instance (I1) and progressing to the following instance by doubling the length $l(\overline{a})$ or both of the lengths $l(\overline{x})$ and $l(\overline{y})$. The GA was run ten times on each instance and the mean runtime $\overline{t}$ in seconds and mean generation count $\overline{g}$ across all runs of that instance was taken. For each collection of runs of an instance we took the standard deviation σ_g of the generation counts and the mean time in seconds taken to compute each population. A summary of results is given in Table 1.

Increasing Rank. These tests were designed to keep the lengths of computed words relatively small while allowing the rank n to increase. We no longer impose the condition of $l(\overline{x}) = l(\overline{y})$. Take s to be the arithmetic mean of the lengths of $\overline{x}$ and $\overline{y}$. Instances were constructed by taking $n = 10, 20$ or 40 and generating random a of maximal length 750, random x and y of maximal length 150 and then reducing the new $b = xay$ to its normal form $\overline{b}$.

We then ran the GA once on each of 505 randomly generated instances for $n = 10$, with 145 instances for $n = 20$ and 52 instances for $n = 40$. We took the time t in seconds to produce a solution and the respective generation count g. The data collected is summarised on Table 2 by grouping the length s of instance into intervals of length fifteen. For example, the range 75–90 means all instances where $s \in [75, 90)$. Across each interval we computed the means $\overline{g}$ and $\overline{t}$ along with the standard deviation σ_g. We now give a brief discussion of the results and some conjectures, and then conclude our work.

5.5   Discussion and Conclusion

Firstly, the mean times given on Tables 1 and 2 depend upon the time complexity of the underlying algebraic operations. We conjecture for $n = 10$ that these have time complexity no greater than $O(k \log k)$ where k is the mean length of all words across the entire run of the GA that we wish to reduce.

Table 1 shows we have a good method for solving large scale problems when the rank is $n = 10$. By Table 2 we observe the GA operates very well in most cases across problems where the mean length of x and y is less than 150 and rank

Table 2. Results of increasing rank from $n = 10$ (upper rows) to $n = 20$ (centre rows) and $n = 40$ (lower rows)

s	15–30	30–45	45–60	60–75	75–90	90–105	105–120	120–135	135–150
$\overline{g}$	227	467	619	965	1120	1740	1673	2057	2412
$\overline{t}$	44	94	123	207	244	384	399	525	652
$\overline{g}$	646	2391	2593	4349	4351	8585	8178	8103	10351
$\overline{t}$	251	897	876	1943	1737	3339	3265	4104	4337
$\overline{g}$	1341	1496	2252	1721	6832	14333	14363	-	-
$\overline{t}$	949	1053	836	1142	5727	10037	11031	-	-

at most forty. Fixing s in a given range, the mean generation count increases at an approximately linearithmic rate as n increases. This seems to hold for all n up to forty, so we conjecture that for a mean instance of problem (P) with given rank n and instance length s the generation count for an average run of the GA lies between $O(sn)$ and $O(sn \log n)$. This conjecture means the GA generation count depends linearly on s (for brevity, we omit the statistical evidence here).

As n increases across the full range of instances of (P), increasing numbers of suicides tend to occur as the GA encounters increasing numbers of local minima. These may be partially explained by observing traceback. For n large, we are likely to have many more blocks than for n small (as the likelihood of two arbitrary generators commuting is larger). While traceback is much more efficient than a purely random method, there are more chances to read δ between blocks. Indeed, there may be so many possible δ that it takes many GA iterations to reduce cost. By experience of this situation, non-asexual methods of reproduction bring the GA out of some local minima. Consider the following typical GA output, where the best chromosomes from populations 44 and 64 (before and after a local minimum) are:

```
Gen 44 (c = 302) : x = 9 6 5 6 7 4 5 -6 7 5 -3 -3 (l = 12)

y = -20 14 12 14 -20 -20 (l = 6)

Gen 64 (c = 300) : x = 9 8 1 7 6 5 6 7 4 5 -6 7 9 5 -3 -3 (l = 16)

y = 14 12 12 -20 14 15 -14 -14 -16 17 15 14 -20 15 -19 -20 -20 -19
-20 18 -17 -16 (l = 22)
```

In this case, cost reduction is not made by a small change in chromosome length, but by a large one. We observe that the cost reduction is made when a chromosome from lower in the ordered population is selected and then mutated, as the new chromosome at population 64 is far longer. In this case it seems traceback acts as a topological sorting method on the generators of the equation E, giving complex systems of cancellation in E which result in a cost deduction greater than one. This suggests that finetuning the parameter set to focus more on reproduction lower in the population and reproduction which causes larger changes in word length may improve performance. Indeed, [3] conjectures that

"It seems plausible to conjecture that sexual mating has the purpose to overcome situations where asexual evolution is stagnant."

Bremermann [3, p. 102]

This implies the GA performs well in comparison to asexual hillclimbing methods. Indeed, this is the case in practice: by making appropriate parameter choices we may simulate such a hillclimb, which experimentally encounters many more local minima. These local minima seem to require substantial changes in the form of χ and ζ (as above); this cannot be done by mere asexual reproduction.

Meanwhile, coupled with a concept of "growing" solutions, we have at least for reasonable values of n an indication of a good underlying deterministic algorithm based on traceback. Indeed, such deterministic algorithms were developed in [2] as the result of analysis of experimental data in our work. This hints that the search space has a "good" structure and may be exploited by appropriately sensitive GAs and other artificial intelligence technologies in our framework.

References

1. R. F. Booth, D. Y. Bormotov, A. V. Borovik, Genetic Algorithms and Equations in Free Groups and Semigroups, Contemp. Math. **349** (2004), 63–80.
2. A. V. Borovik, E. S. Esyp, I. V. Kazatchkov, V. N. Remeslennikov, Divisibility Theory and Complexity of Algorithms for Free Partially Commutative Groups, Contemp. Math. **378** (Groups, Languages, Algorithms), 2005.
3. H. J. Bremermann, Optimization Through Evolution and Recombination, Self-Organizing Systems (M. C. Yovits et al., eds.), Washington, Spartan Books (1962), 93–106.
4. J. Holland, Adaptation in Natural and Artificial Systems (5th printing), MIT Press, Cambridge, Massachusetts, 1998.
5. K. -H. Ko, Braid Group and Cryptography, 19th SECANTS, Oxford, 2002.
6. D. Knuth, P. Bendix, Simple Word Problems in Universal Algebra, Computational Problems in Abstract Algebras (J. Leech, ed.), Pergamon Press 1970, 263–297.
7. A. D. Miasnikov, Genetic Algorithms and the Andrews-Curtis Conjecture, Internat. J. Algebra Comput. **9** (1999), no. 6, 671–686.
8. A. D. Miasnikov, A. G. Myasnikov, Whitehead Method and Genetic Algorithms, Contemp. Math. **349** (2004), 89–114.
9. Z. Michalewicz, Genetic Algorithms + Data Structures = Evolution Programs (3rd rev. and extended ed.), Springer-Verlag, Berlin, 1996.
10. L. VanWyk, Graph Groups are Biautomatic, J. Pure Appl. Algebra **94** (1994), no. 3, 341–352.
11. A. Vershik, S. Nechaev, R. Bikbov, Statistical Properties of Braid Groups in Locally Free Approximation, Comm. Math. Phys. **212** (2000), 59–128.
12. C. Wrathall, The Word Problem for Free Partially Commutative Groups, J. Symbolic Comp. **6** (1988), 99–104.
13. C. Wrathall, Free partially commutative groups, Combinatorics, Computing and Complexity (Tianjing and Beijing, 1988) 195–216, Math. Appl (Chin. Ser. 1) Kluwer Acad. Publ., Dordrecht, 1989.

A GRASP and Branch-and-Bound Metaheuristic for the Job-Shop Scheduling

Susana Fernandes and Helena R. Lourenço

Universidade do Algarve, Faro, Portugal
sfer@ualg.pt
Universitat Pompeu Fabra, Barcelona, Spain
helena.ramalhinho@upf.edu

Abstract. This paper presents a simple algorithm for the job shop scheduling problem that combines the local search heuristic GRASP with a branch-and-bound exact method of integer programming. The proposed method is compared with similar approaches and leads to better results in terms of solution quality and computing times.

1 Introduction

The job-shop scheduling problem has been known to the operations research community since the early 50's [16]. It is considered a particularly hard combinatorial optimization problem of the NP-hard class [15] and it has numerous practical applications; which makes it an excellent test problem for the quality of new scheduling algorithms. These are main reasons for the vast bibliography on both exact and heuristic procedures applied to this scheduling problem. The paper of Jain and Meeran [16] includes an exhaustive survey not only of the evolution of the definition of the problem, but also of all the techniques applied to it.

Recently a new class of procedures that combine local search based (meta) heuristics and exact algorithms have been developed. Fernandes and Lourenço [13] designated these methods by Optimized Search Heuristics (OSH).

In this paper we present a simple OSH procedure for the job-shop scheduling problem that combines a GRASP algorithm with a branch-and-bound method.

We first introduce the job-shop scheduling problem. We present a short review of existent OSH methods applied to this problem and proceed describing the procedure developed. Computational results are presented along with comparisons to other procedures.

2 The Job-Shop Scheduling Problem

The job-shop scheduling problem considers a set of jobs to be processed on a set of machines. Each job is defined by an ordered set of operations and each operation is assigned to a machine with a predefined constant processing time (preemption is not allowed). The order of the operations within the jobs and its correspondent machines

C. Cotta and J. van Hemert (Eds.): EvoCOP 2007, LNCS 4446, pp. 60–71, 2007.
© Springer-Verlag Berlin Heidelberg 2007

are fixed a priori and independent from job to job. To solve the problem we need to find a sequence of operations on each machine respecting some constraints and optimizing some objective function. It is assumed that two consecutive operations of the same job are assigned to different machines, each machine can only process one operation at a time and that different machines can not process the same job simultaneously. We will adopt the maximum of the completion time of all jobs – the makespan – as the objective function.

Formally let $O = \{0,\ldots,o+1\}$ be the set of operations with 0 and $o+1$ being the dummy operations representing the start and end of all jobs, respectively. Let M be the set of machines, A the set of arcs between consecutive operations of each job and E_k the set of all possible pairs of operations processed by machine k, with $k \in M$. We define $p_i > 0$ as the constant processing time of operation i and t_i is the variable representing the start time of operation i. The following mathematical formulation for the job shop scheduling problem is widely used:

The constraints in (1) state the precedences of operations within jobs and also that no two operations of the same job can be processed simultaneously (because $p_i > 0$). Expressions (3) are named "capacity constraints" and assure there are no overlaps of operations on the machines.

A common representation for the job-shop problem is the disjunctive graph $G = (O, A, E)$ [22]; where O is the node set, corresponding to the set of operations; A is the set of arcs between consecutive operations of the same job, and E is the set of edges between operations processed by the same machine. For every node j of $O/\{0, o+1\}$ there are unique nodes i and l such that arcs (i, j) and (j, l) are elements of A. Node i is called the job predecessor of node j - $jp(j)$ and l is the job successor of j - $js(j)$. Finding a solution to the job-shop scheduling problem means replacing every edge of the respective graph with a directed arc, constructing an acyclic directed graph $D_S = (O, A \cup S)$ where $S = \bigcup_k S_k$ corresponds to an acyclic union of sequences of operations for each machine k. The optimal solution is the one represented by the graph D_S having the critical path from 0 to $o+1$ with the smallest length or makespan.

$(JSSP)$

$$\min t_{o+1}$$

s.t.
$$t_j - t_i \geq p_i \qquad\qquad (i, j) \in A \qquad\qquad (1)$$
$$t_i \geq 0 \qquad\qquad i \in O \qquad\qquad (2)$$
$$t_j - t_i \geq p_i \vee t_i - t_j \geq p_j \qquad (i, j) \in E_k, k \in M \quad (3)$$

Fig. 1. Mathematical formulation for the Job-Shop Problem

3 Review of Optimized Search Heuristics

In the literature we can find a few works combining metaheuristics with exact algorithms applied to the job shop scheduling problem, designated as Optimized Search Heuristics (OSH) by Fernandes and Lourenço [13]. Different combinations of different procedures are present in the literature, and there are several applications of the OSH methods to different problems.

Chen et al. [8] and Denzinger and Offermann [11] design parallel algorithms that use asynchronous agents information to build solutions; some of these agents are genetic algorithms, others are branch-and-bound algorithms.

Tamura et al. [27] design a genetic algorithm where the fitness of each individual, whose chromosomes represent each variable of the integer programming formulation, is the bound obtained solving Lagrangian relaxations.

The works [1], [3], [7] and [4] use an exact algorithm to solve a sub problem within a local search heuristic for the job shop scheduling. Caseau and Laburthe [7] build a local search where the neighborhood structure is defined by a subproblem that is solved exactly using constraint programming.

Applegate and Cook [3] develop the shuffle heuristic. At each step of the local search the processing orders of the jobs on a small number of machines is fixed, and a branch-and-bound algorithm completes the schedule. The shifting bottleneck heuristic, due to Adams, Balas and Zawack [1], is an iterated local search with a construction heuristic that uses a branch-and-bound to solve the subproblems of one machine with release and due dates. Balas and Vazacopoulos [4] work with the shifting bottleneck heuristic and design a guided local search, over a tree search structure, that reconstructs partially destroyed solutions.

Lourenço [18] and Lourenço and Zwijnenburg [19] use branch-and-bound algorithms to strategically guide an iterated local search and a tabu search algorithm. The diversification of the search is achieved applying a branch-and-bound method to solve a one-machine scheduling problem subproblem obtained from the incumbent solution.

The interesting work done by Danna, Rothberg and Le Pape [9] "applies the spirit of metaheuristics" in an exact algorithm. Within each node of a branch-and-cut tree, the solution of the linear relaxation is used to define the neighborhood of the current best feasible solution. The local search consists in solving the restricted MIP problem defined by the neighborhood.

4 GRASP and Branch-and-Bound

We developed a simple optimized search heuristic that combines a GRASP algorithm with a branch-and-bound method. The branch-and-bound method is used within the GRASP to solve subproblems of one machine scheduling.

GRASP [12] is an iterative process where each iteration consists of two steps: a randomized building step of a greedy nature and a local search step. At the building phase, a solution is constructed joining one element at a time. Each element is evaluated by a greedy function and incorporated (or not) in a restricted candidate list (RCL). The element to join the solution is chosen randomly from the RCL. Each time a new element is added to the partial solution the algorithm proceeds with the local search step and the local optimum updates the current solution. The all process is repeated until the solution is complete.

4.1 Building Step

We define the sequence of operations at each machine as the elements to join the solution, and the makespan as the greedy function to evaluate them. In order to build a restricted candidate list of this elements (RCL), we solve exactly all the one machine problems and identify the best $(\underline{f})$ and worst $(\overline{f})$ makespans. A machine k is included in the RCL if $f(x_k) \geq \overline{f} - \alpha(\overline{f} - \underline{f})$, where $f(x_k)$ is the makespan of machine k and α is a uniform random number in $(0,1)$. This semi-greedy randomised procedure is biased towards the machine with the higher makespan, the bottleneck machine, in the sense that machines with low values of makespan have less probability of being included in the restricted candidate list.

To solve the one machine scheduling problems we use the branch-and-bound algorithm of Carlier [6]. The objective function of the algorithm is to minimize the completion time of all jobs. This one-machine scheduling problem considers that, associated to each job j, there are the following values (obtained from the current solution): the processing time (p_j), a release date (r_j) and an amount of time (q_j) that the job stays in the system after being processed.

At each node of the branch-and-bound tree the upper bound is computed using the algorithm of Schrage [23]. This algorithm gives priority to higher values of the tails (q_j) when scheduling released jobs. We break ties preferring jobs with larger processing times.

The lower bound is computed as in [6]. The value of the solution where preemption is allowed, is used to strengthen this lower bound. We introduce a slight modification, forcing the lower bound of a node never to be smaller than the one of its father in the tree.

At the first iteration we consider the graph $D = (O, A)$ (without the edges connecting operations that share the same machine) to compute release dates and tails. Incorporating a new machine in the solution means adding to the graph the arcs representing the sequence of operations in that machine. In terms of the mathematical formulation, this means choosing one of the inequalities of the disjunctive constraints (3) correspondent to the machine. We then update the makespan of the partial solution and the release dates and tails of unscheduled operations using the algorithm of Taillard [26].

4.2 Local Search

In order to build a simple local search algorithm we need to design a neighborhood structure, the way to inspect the neighborhood of a given solution, and a procedure to evaluate the quality of each solution.

We use a neighborhood structure very similar to the NB neighborhood of Dell'Amico and Trubian [10] and the one of Balas and Vazacopoulos [4]. To describe the moves that define this neighborhood we use the notion of blocks of critical operations. A block of critical operations is a maximal ordered set of consecutive operations of a critical path, sharing the same machine. Borrowing the nomination of Balas and Vazacopoulos [4] we speak of forward and backward moves over forward and backward critical pairs of operations. Let $L(i, j)$ denote the length of the critical path from node i to node j.

Two operations u and v form a forward critical pair (u, v) if: a) they both belong to the same block; b) v is the last operation of the block; c) operation $js(v)$ also belongs to the same critical path; d) the length of the critical path from v to $o+1$ is not less than the length of the critical path from $js(u)$ to $o+1$ ($L(v, o+1) \geq L(js(u), o+1)$).

Two operations u and v form a backward critical pair (u, v) if: a) they both belong to the same block; b) u is the first operation of the block; c) operation $jp(u)$ also belongs to the same critical path; d) the length of the critical path from 0 to u, including the processing time of u, is not less than the length of the critical path from 0 to $jp(v)$, including the processing time of $jp(v)$ ($L(0, u) + p_u \geq L(0, jp(v)) + p_{jp(v)}$)).

Conditions d) guarantee that all moves lead to feasible solutions [4].

A forward move is executed by moving operation u to be processed immediately after operation v. A backward move is executed by moving operation v to be processed immediately before operation u.

When inspecting the neighborhood ($N(x, M_k)$) of a given solution x with M_k machines already scheduled, we stop whenever we find a neighbor with a best evaluation value than the makespan of x.

To evaluate the quality of a neighbor of a solution x, produced by a move over a critical pair (u, v), we need only to compute the length of all the longest paths through the operations that were between u and v in the critical path of solution x. This evaluation is computed using the algorithm described in [4].

4.3 GRASP_B&B

Let *runs* be the total number of runs, M the set of machines and $f(x)$ the makespan of a solution x. The procedure GRASP_B&B can be generally described by the pseudo-code in Fig. 2:

```
GRASP_B&B (runs)
(1)        M := {1,···,m}
(2)        for r = 1 to runs
(3)        x := { }
(4)        K := M
(5)        while K ≠ { }
(6)            foreach k ∈ K
(7)                        xₖ := CARLIER_B&B(k)
(8)            k* := SEMIGREEDY(K)
(9)            x := x ∪ xₖ*
(10)           f(x) := TAILLARD(x)
(11)           K := K \ {k*}
(12)           if |K| < |M| - 1
(13)               x := LOCALSEARCH(x, M \ K)
(14)           if x* not initialized or f(x) < f*
(15)               x* := x
(16)               f* := f(x)
(17)       return x*
```

Fig. 2. Pseudo-code of algorithm GRASP_B&B

5 Computational Results

We have tested the algorithm GRASP_B&B (coded in C) on a Pentium 4 CPU 2.80 GHz, on the benchmark instances abz5-9 [1], ft6, ft10, ft20 [14], la01-40 [17],

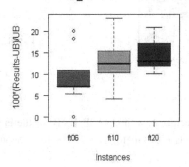

GRASP_B&B: % from best UB

Fig. 3. Boxplots of RE_{UB} achieved with GRASP_B&B for the **ft** instances. ft06: (6*6); ft10: (10*10); ft20: (20*5).

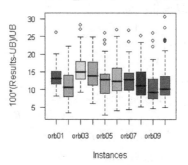

Fig. 4. Boxplots of RE_{UB} achieved with GRASP_B&B for the **orb** instances. orb01-10: (10*10).

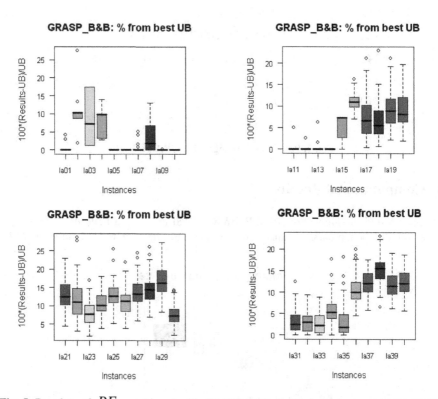

Fig. 5. Boxplots of RE_{UB} achieved with GRASP_B&B for the **la** instances. la01-05: (10*5); la06-10: (15*5); la11-15: (20*5); la16-20: (10*10); la21-25: (15*10); la26-30: (20*10); la31-35: (30*10); la36-40: (15*15).

orb01-10 [3], swv01-20 [24], ta01-70 [25] and yn1-4 [28]. The dimension of each instance is defined as the number of jobs times the number of machines ($n*m$).

Because of space limitations, in this work we will only present the results for instances ft6, ft10, ft20, la01-40 and orb01-10.

We show the results of running the algorithm 100 times for each instance presenting boxplots (figures 3 – 5) of RE_{UB}, the percentage of relative error to the best known upper bound (UB), calculated as follows:

$$RE_{UB}(x) = 100\% \times \frac{f(x) - UB}{UB}$$

We gathered the values of the upper bounds from [16], [20] and [21].

The boxplots show that the quality achieved is more dependent on the ratio n/m than on the absolute numbers of jobs and machines. There is no big dispersion of the solution values achieved by the algorithm in the 100 runs executed, except maybe for instance la3. The number of times the algorithm achieves the best values reported is high enough, so these values are not considered outliers of the distribution of the results, except for instances ft06 and la38. On the other end, the worse values occur very seldom and are outliers for the majority of the instances.

Although this is a very simple (and fast) algorithm, the best values are not worse than the best known upper bound for 22 of the 152 instances used in this study.

5.1 Comparison to Other Algorithms

GRASP_B&B is a simple GRASP algorithm with a construction phase very similar to the one of the shifting bottleneck procedure. Therefore we show comparative results to two other procedures designed for the job shop problem; a simple GRASP procedure of Binato et al. [5] and the shifting bottleneck procedure of Adams et al. [1].

The building block of the construction phase of the GRASP in [5] is a single operation of a job. In their computational results, they present the time in seconds per thousand iterations (an iteration is one building phase followed by a local search) and the thousands of iterations. For a comparison purpose we multiply these values to get the total computation time. For GRASP_B&B we present the time to the best solution found (btime) and the total time of all runs (ttime), in seconds. As the tables show, our algorithm is much faster. Whenever our GRASP_B&B achieves a solution not worse than theirs, we present the respective value in bold. This happens for 25 of the 53 instances whose results where compared.

Table 1. Comparing GRASP_B&B with (Binato et al 2001) and (Adams et al. 1988) - **ft** instances

name	GRASP_B&B	btime(s)	ttime (s)	GRASP	time (s)	Shifting Bottleneck	time (s)
ft06	**55**	0.1274	0.1400	55	70	55	1.5
ft10	970	0.5800	1.0000	938	261290	1015	10.1
ft20	1283	0.0094	0.4690	1169	387430	1290	3.5

Table 2. Comparing GRASP_B&B with (Binato et al 2001) and (Adams et al. 1988) - **la** instances

name	GRASP_B&B	btime (s)	ttime (s)	GRASP	time (s)	Shifting Bottleneck	time (s)
la01	**666**	0.0017	0.1720	666	140	666	1.26
la02	667	0.0437	0.1560	655	140	720	1.69
la03	605	0.0066	0.2190	604	65130	623	2.46
la04	607	0.0051	0.1710	590	130	597	2.79
la05	**593**	0.0011	0.1100	593	130	593	0.52
la06	**926**	0.0017	0.1710	926	240	926	1.28
la07	**890**	0.002	0.2030	890	250	890	1.51
la08	**863**	0.0149	0.2970	863	240	868	2.41
la09	**951**	0.0028	0.2810	951	290	951	0.85
la10	**958**	0.0014	0.1410	958	250	959	0.81
la11	**1222**	0.0027	0.2660	1222	410	1222	2.03
la12	**1039**	0.0027	0.2650	1039	390	1039	0.87
la13	**1150**	0.0038	0.3750	1150	430	1150	1.23
la14	**1292**	0.0022	0.2180	1292	390	1292	0.94
la15	**1207**	0.0453	0.9060	1207	410	1207	3.09
la16	1012	0.0221	0.7350	946	155310	1021	6.48
la17	787	0.0843	0.7660	784	60300	796	4.58
la18	854	0.3	0.7500	848	58290	891	10.2
la19	861	0.4554	0.9690	842	31310	875	7.4
la20	920	0.0813	0.8130	907	160320	924	10.2
la21	1092	0.1023	2.0460	1091	325650	1172	21.9
la22	**955**	0.9884	1.7970	960	315630	1040	19.2
la23	1049	1.7388	1.8900	1032	65650	1061	24.6
la24	**971**	0.627	1.8440	978	64640	1000	25.5
la25	**1027**	0.5388	1.7960	1028	64640	1048	27.9
la26	**1265**	3.0375	3.3750	1271	109080	1304	48.5
la27	**1308**	0.1781	3.5620	1320	110090	1325	45.5
la28	1301	0.15	3.0000	1293	110090	1256	28.5
la29	**1248**	0.857	3.2960	1293	112110	1294	48
la30	1382	0.8653	3.3280	1368	106050	1403	37.8
la31	**1784**	0.0702	7.0160	1784	231290	1784	38.3
la32	**1850**	0.5612	6.2350	1850	241390	1850	29.1
la33	**1719**	1.265	7.9060	1719	241390	1719	25.6
la34	**1721**	3.8093	8.2810	1753	240380	1721	27.6
la35	**1888**	0.2844	5.6880	1888	222200	1888	21.3
la36	**1325**	0.0853	4.2650	1334	115360	1351	46.9
la37	1479	4.0295	4.7970	1457	115360	1485	6104
la38	1274	0.7153	5.1090	1267	118720	1280	57.5
la39	1309	2.9835	4.4530	1290	115360	1321	71.8
la40	1291	3.5581	5.3910	1259	123200	1326	76.7

Table 3. Comparing GRASP_B&B with (Binato et al 2001) - **orb** instances

name	GRASP_B&B	btime (s)	ttime (s)	GRASP	time (s)
orb01	1145	0.0296	0.9850	1070	116290
orb02	918	0.0953	0.9530	889	152380
orb03	1098	0.335	1.0150	1021	124310
orb04	1066	0.8213	1.1250	1031	124310
orb05	911	0.105	0.8750	891	112280
orb06	1050	0.4812	1.0460	1013	124310
orb07	414	0.2764	1.0630	397	128320
orb08	945	0.3093	1.0310	909	124310
orb09	978	0.2809	0.9060	945	124310
orb10	991	0.2276	0.8430	953	116290

The comparison between the shifting bottleneck procedure [1] and the GRASP_B&B is presented in tables 1 and 2. Comparing the computation times of both procedures, our GRASP is slightly faster than the shifting bottleneck for smaller instances. Given the distinct computers used in the experiments we would say that this is not meaningful, but the difference does get accentuated as the dimensions grow. Whenever GRASP_B&B achieves a solution better than the shifting bottleneck procedure, we present the respective value underlined. This happens in 25 of the 43 instances whose results where compared, and in 16 of the remaining 18 instances the best value found was the same.

6 Conclusions

We have designed a very simple optimized search heuristic, the GRASP_B&B. It is intended to be a starting point for a more elaborated metaheuristic. We have compared it to other base procedures used within more complex algorithms; namely a GRASP [5], which is the base for a GRASP with path-relinking procedure [2], and the shifting bottleneck procedure, incorporated in the successful guided local search [4]. The comparison to the GRASP [5] shows that our procedure is much faster than theirs. The quality of their best solution is slightly better than ours in 60% of the instances tested. When comparing GRASP_B&B with the shifting bottleneck, ours is still faster, and it achieves better solutions, except for 2 of the comparable instances.

Acknowledgement

Susana Fernandes' work is suported by the the programm POCI2010 of the Portuguese Fundação para a Ciência e Tecnologia. Helena R. Lourenço's work is supported by Ministerio de Educación y Ciencia, Spain, SEC2003-01991/ECO.

References

1. Adams, J., E. Balas and D. Zawack (1988). "The Shifting Bottleneck Procedure for Job Shop Scheduling." Management Science, vol. 34(3): pp. 391-401.
2. Aiex, R. M., S. Binato and M. G. C. Resende (2003). "Parallel GRASP with path-relinking for job shop scheduling." Parallel Computing, vol. 29(4): pp. 393-430.
3. Applegate, D. and W. Cook (1991). "A Computational Study of the Job-Shop Scheduling Problem." ORSA Journal on Computing, vol. 3(2): pp. 149-156.
4. Balas, E. and A. Vazacopoulos (1998). "Guided Local Search with Shifting Bottleneck for Job Shop Scheduling." Management Science, vol. 44(2): pp. 262-275.
5. Binato, S., W. J. Hery, D. M. Loewenstern and M. G. C. Resende (2001). "A GRASP for Job Shop Scheduling." In C.C. Ribeiro and P. Hansen, editors, Essays and surveys on metaheuristics, pp. 59-79. Kluwer Academic Publishers.
6. Carlier, J. (1982). "The one-machine sequencing problem." European Journal of Operational Research, vol. 11: pp. 42-47.
7. Caseau, Y. and F. Laburthe (1995), "Disjunctive scheduling with task intervals", Technical Report LIENS, 95-25, Ecole Normale Superieure Paris.
8. Chen, S., S. Talukdar and N. Sadeh (1993). "Job-shop-scheduling by a team of asynchronous agentes", Proceedings of the IJCAI-93 Workshop on Knowledge-Based Production, Scheduling and Control. Chambery France.
9. Danna, E., E. Rothberg and C. L. Pape (2005). "Exploring relaxation induced neighborhoods to improve MIP solutions." Mathematical Programming, Ser. A, vol. 102: pp. 71-90.
10. Dell'Amico, M. and M. Trubian (1993). "Applying Tabu-Search to the Job-Shop Scheduling Problem."
11. Denzinger, J. and T. Offermann (1999). "On Cooperation between Evolutionary Algorithms and other Search Paradigms", Proceedings of the 1999 Congress on Evolutionary Computational.
12. Feo, T. and M. Resende (1995). "Greedy Randomized Adaptive Search Procedures." Journal of Global Optimization, vol. 6: pp. 109-133.
13. Fernandes, S. and H.R. Lourenço (2006), "Optimized Search methods", Working paper, Universitat Pompeu Fabra, Barcelona, Spain.
14. Fisher, H. and G. L. Thompson (1963). Probabilistic learning combinations of local job-shop scheduling rules. In J. F. Muth and G. L. Thompson eds. Industrial Scheduling. pp. 225-251. Prentice Hall, Englewood Cliffs.
15. Garey, M. R. and D. S. Johnson (1979). Computers and Intractability: A Guide to the Theory of NP-Completeness. San Francisco, Freeman.
16. Jain, A. S. and S. Meeran (1999). "Deterministic job shop scheduling: Past, present and future." European Journal of Operational Research, vol. 133: pp. 390-434.
17. Lawrence, S. (1984), "Resource Constrained Project Scheduling: an Experimental Investigation of Heuristic Scheduling techniques", Graduate School of Industrial Administration, Carnegie-Mellon University.
18. Lourenço, H. R. (1995). "Job-shop scheduling: Computational study of local search and large-step optimization methods." European Journal of Operational Research, vol. 83: pp. 347-367.
19. Lourenço, H. R. and M. Zwijnenburg (1996). Combining large-step optimization with tabu-search: Application to the job-shop scheduling problem. In I. H. Osman and J. P. Kelly eds. Meta-heuristics: Theory & Applications. Kluwer Academic Publishers.

20. Nowicki, E. and C. Smutnicki (2005). "An Advanced Tabu Search Algorithm for the Job Shop Problem." Journal of Scheduling, vol. 8: pp. 145-159.
21. Nowicki, E. and C. Smutniki (1996). "A Fast Taboo Search Algorithm for the Job Shop Problem." Management Science, vol. 42(6): pp. 797-813.
22. Roy, B. and B. Sussman (1964), "Les probèms d'ordonnancement avec constraintes disjonctives", Note DS 9 bis, SEMA, Paris.
23. Schrage, L. (1970). "Solving resource-constrained network problems by implicit enumeration: Non pre-emptive case." Operations Research, vol. 18: pp. 263-278.
24. Storer, R. H., S. D. Wu and R. Vaccari (1992). "New search spaces for sequencing problems with application to job shop scheduling." Management Science, vol. 38(10): pp. 1495-1509.
25. Taillard, E. D. (1993). "Benchmarks for Basic Scheduling Problems." European Journal of Operational Research, vol. 64(2): pp. 278-285.
26. Taillard, É. D. (1994). "Parallel Taboo Search Techniques for the Job Shop Scheduling Problem." ORSA Journal on Computing, vol. 6(2): pp. 108-117.
27. Tamura, H., A. Hirahara, I. Hatono and M. Umano (1994). "An approximate solution method for combinatorial optimisation." Transactions of the Society of Instrument and Control Engineers, vol. 130: pp. 329-336.
28. Yamada, T. and R. Nakano (1992). A genetic algorithm applicable to large-scale job-shop problems. In R. Manner and B. Manderick eds. Parallel Problem Solving from Nature 2. pp. 281-290. Elsevier Science.

Reducing the Size of Traveling Salesman Problem Instances by Fixing Edges

Thomas Fischer and Peter Merz

Distributed Algorithms Group
Department of Computer Science
University of Kaiserslautern, Germany
{fischer,pmerz}@informatik.uni-kl.de

Abstract. The Traveling Salesman Problem (TSP) is a well-known NP-hard combinatorial optimization problem, for which a large variety of evolutionary algorithms are known. However, these heuristics fail to find solutions for large instances due to time and memory constraints. Here, we discuss a set of edge fixing heuristics to transform large TSP problems into smaller problems, which can be solved easily with existing algorithms. We argue, that after expanding a reduced tour back to the original instance, the result is nearly as good as applying the used solver to the original problem instance, but requiring significantly less time to be achieved. We claim that with these reductions, very large TSP instances can be tackled with current state-of-the-art evolutionary local search heuristics.

1   Introduction

The Traveling Salesman Problem (TSP) is a widely studied combinatorial optimization problem, which is known to be NP-hard [1].

Let $G = (V, E, d)$ be an edge-weighted, directed graph, where V is the set of $n = |V|$ vertices, $E \subseteq V \times V$ the set of of (directed) edges and $d : E \to \mathbb{R}^+$ a *distance function* assigning each edge $e \in E$ a distance $d(e)$. A *path* is a list $(u_1, \ldots, u_k)$ of vertices $u_i \in V$ ($i = 1, \ldots k$) holding $(u_i, u_{i+1}) \in E$ for $i = 1, \ldots, k - 1$. A *Hamiltonian cycle* in G is a path $p = (u_1, \ldots, u_k, u_1)$ in G, where $k = n$ and $\bigcup_{i=1}^{k} u_i = V$ (each vertex is visited exactly once except for u_1). The TSP's objective is to find a Hamiltonian cycle t for G that minimizes the cost function $C(t) = \sum_{i=1}^{k-1} d((u_i, u_{i+1})) + d((u_k, u_1))$ (weights of the edges in t added up). Depending on the distance function d, a TSP instance may be either *symmetric* (for all $u_1, u_2 \in V$ holds $d((u_1, u_2)) = d(u_2, u_1))$) or *asymmetric* (otherwise). Most applications and benchmark problems are *Euclidean*, i. e., the vertices V correspond to points in an Euclidean space (mostly 2-dimensional) and the distance function represents an Euclidean distance metric. The following discussion focuses on symmetric, Euclidean problem instances.

Different types of algorithms for the TSP are known, such as exact algorithms [2,3] or local search algorithms [4]. Among the best performing algorithms are those utilizing Lin-Kernighan local search within an evolutionary framework such

C. Cotta and J. van Hemert (Eds.): EvoCOP 2007, LNCS 4446, pp. 72–83, 2007.
© Springer-Verlag Berlin Heidelberg 2007

as Iterated Lin-Kernighan [5] or memetic algorithms [6,7]. Even exact algorithms like Branch & Cut rely on these heuristics. The heuristics used in Concorde [8] to find near-optimum solutions to large TSP instances is essentially a memetic algorithm using the Lin-Kernighan (LK) heuristics as local search and tour-merging [9] for recombination [10]. As the TSP is NP-hard, computation time is expected to grow exponentially with the instance size. E. g. for a TSP instance with 24 978 cities, even sophisticated heuristic algorithms such as Helsgaun's LK (LK-H) [11] require several hours to find solutions within 1% distance to the optimum. For the same instance, an exact algorithm required 85 CPU years to prove a known tour's optimality [12].

The problem of time consumption can be approached by distributing the computation among a set of computers using distributed evolutionary algorithms (DEA) [13,14]. Another problem when solving extremely large TSP instances such as the World TSP [15] is an algorithm's memory consumption, as data structures such as neighbor or candidate lists have to be maintained. We address this problem in this paper by proposing different *edge fixing heuristics*, which may reduce the problem to a size suitable for standard TSP solvers. In the general fixing scheme heuristics select edges of an existing tour for fixing; paths of fixed edges are merged into a single fixed edge reducing the instance size. Fixed edges are 'tabu' for the TSP solver, which is applied to the reduced instance in a second step. Finally, the optimized tour is expanded back to a valid solution for the original problem by releasing fixed edges and paths.

The remainder of this section discusses related work from Walshaw. In Sect. 2 problem reduction techniques based on fixing edges are discussed. A set of TSP instances is analyzed in Sect. 3 regarding the discussed fixing heuristics. Sect. 4 discusses the results when applying the fixing heuristics to an evolutionary local search. Sect. 5 summarizes our findings.

1.1 Related Work

Only limited research regarding the reduction of TSP instances in relation with evolutionary local search has been done. The primary related work to our concept is the *multilevel approach* by Walshaw [16], which has been applied to several graph problems including the TSP [17]. Basically, multilevel algorithms work as follows: In the first phase a given graph is recursively coarsened by matching and merging node pairs generating smaller graphs at each level. The coarsening stops with a minimum size graph, for which an optimal solution can easily be found. In the second phase, the recursion backtracks, uncoarsening each intermediate graph and finally resulting in a valid solution of the original problem. In each uncoarsening step, the current solution is refined by some optimization algorithm. It has been reported that this strategy results in better solutions compared to applying the optimization algorithm to the original graph only. When uncoarsening again, the optimization algorithm can improve the current level's solution based on an already good solution found in the previous level. As the coarsening step defines the solution space of a recursion level, its strategy is decisive for the quality of the multilevel algorithm.

In [17] the multilevel approach has been applied to the TSP using a CLK algorithm [18] for optimization. Here, a multilevel variant (MLC$^{N/10}$LK) of CLK gains better results than the unmodified CLK, being nearly 4 times faster. The coarsening heuristics applied to the TSP's graph matches node pairs by adding a fixed edge in between. In each step, nodes are selected and matched with their nearest neighbor, if feasible. Nodes involved in an (unsuccessful) matching may not be used in another matching at the same recursion level to prevent the generation of sub-tours. Recursion stops when only two nodes and one connecting edge are left.

2   Problem Reduction by Fixing Edges

To reduce a TSP instance's size different approaches can be taken. Approaches can be either node-based or edge-based. At a different level, approaches can be based only on a TSP instance or using an existing solution, respectively.

A *node-based approach* may work as follows: Subsets of nodes can be merged into meta-nodes (cluster) thus generating a smaller TSP instance. Within a meta-node a cost-effective path connecting all nodes has to be found. The path's end nodes will be connected to the edges connecting the meta-node to its neighbor nodes building a tour through all meta-nodes. Problems for this approach are (i) how to group nodes into meta-nodes (ii) how to define distances between meta-nodes (iii) which two nodes of a cluster will have outbound edges. In an *edge-based approach*, a sequence of edges can be merged into a meta-edge, called a fixed path. Subsequently, the inner edges and nodes are no longer visible and this meta-edge has to occur in every valid tour for this instance. Compared to the node-based approach, problems (ii) and (iii) do not apply, as the original node distances are still valid and a fixed path has exactly two nodes with outbound edges. So, the central problem is how to select edges merged into a meta-edge. Examples for both node-based and edge-based problem reductions are shown in Fig. 1.

Edges selected for merging into meta-edges may be chosen based on instance information only or on a tour's structure. The former approach may select from an instance with n nodes any of the $\frac{n(n-1)}{2}$ edges for a merging step, the latter approach reuses only edges from a given tour (n edges). The *tour-based approach*'s advantage is a smaller search space and the reuse of an existing tour's inherent knowledge. Additionally, this approach can easily be integrated into memetic algorithms. A disadvantage is that the restriction to tour edges will limit the fixing effect especially in early stages of a local search when the tour quality is not sufficient.

Walshaw's multilevel TSP approach focuses on an edge-based approach considering the TSP instance only. In this paper, we will discuss edge-based approaches, too, but focus on the following tour-based edge fixing heuristics:

Minimum Spanning Tree (MST). Tour edges get fixed when they occur in a *minimum spanning tree* (MST) for the tour's instance. This can be motivated by the affinity between the TSP and the MST problem [19], as the latter

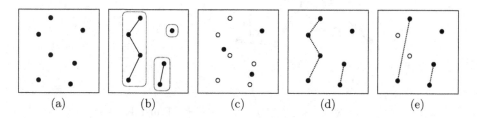

Fig. 1. Examples of node-based and edge-based problem reductions. Starting from the original problem instance (a), the node-based approach assigns node sets to clusters (marked by dashed boxes) and defines a spanning path within each cluster (b). Subsequently, in (c) only representatives of the clusters have to be considered (black nodes, here arbitrarily located at each cluster's center), whereas the original nodes (white) can be ignored. For the edge-based approach, edges to be fixed have to be selected (dotted lines in (d)). Subsequently, paths can be merged to single edges and inner path nodes (white) may be ignored (e).

one can be used to establish a lower bound for the TSP. However, global instance knowledge in form of an MST (complexity of $O(m + n \log n)$ for m edges using Fibonacci heaps) has to be available in advance.

Nearest Neighbor (NN). As already exploited by the *nearest neighbor* tour construction heuristics, edges between a node and it's nearest neighbor are likely to occur in optimal tours thus being promising fixing candidates, too. Determining nearest neighbor lists may be computationally expensive (complexity of $O(n^2 \log n)$), but can be sped up e. g. by using kd-trees [20,21].

Lighter than Median (<M). Edges that length is below the median over all edges' lengths in a tour are selected, as it is beneficial to keep short edges by fixing them and leaving longer edges for further optimization. The most expensive operation of this approach is the necessary sorting of all tour edges (complexity of $O(n \log n)$). There may be tours that have very few different edge lengths resulting in a small number of edges that are strictly shorter than the median.

Close Pairs (CP). Here, a tour edge's length is compared to the lengths of the two neighboring edges. The edge will be fixed if it is shorter than both neighboring edges and the edge's nodes therefore form a *close pair*. This approach considers only local knowledge (edge and its two neighbor edges) allowing it to be applied even on large instances. It is expected to work well in graphs with both sparse and dense regions.

The actual number of edges selected by one of the above heuristics during a fixing step depends on the current tour and the problem instance. For the first two heuristics (MST and NN) it can be expected that more edges will be selected with better tours. For the lighter than median variant (<M) and the close pairs variant (CP) at most half of all edges may be selected. In the example in Fig. 2, solid lines represent tour edges and dotted lines represent edges of the minimal spanning tree (MST). Out of 7 tour edges, 5 edges are MST edges, too, 4 edges

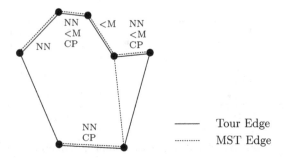

Fig. 2. Example tour and minimum spanning tree with different edge properties highlighted. Tour and MST edges are drawn in different line styles, properties *nearest neighbor* (NN), *lighter than median* (<M), and *close pairs* (CP) are shown as markings next to the edges.

connect nearest neighbors (NN), 3 edges connect close node pairs (CP) and 3 edges are lighter than the edge weight median (<M).

For asymmetric TSP (ATSP) instances edge fixation heuristics can be developed, too, but this approach has not been pursued here. Applying the fixations heuristics presented here to ATSP instances poses new questions such as determining the direction when the fixation heuristics is based on undirected knowledge (e. g. from an MST).

3   Analysis of TSP Instances

The tests in this section were performed to evaluate if the edge selection strategies discussed in this paper are applicable for edge-fixation heuristics. Probabilities of each selection strategy to selected a tour edge and the probabilities if the selected edge is part of an optimal tour were evaluated. These criteria describe the quantity and quality, respectively, of a selection strategy.

For our analysis, five TSP instances have been selected: From the TSPLIB collection [22] instances brd14051, d15112, and d18512 and from a collection of national TSPs [23] instances it16862, and sw24978 were taken (numbers in the instances' names denote the problem size). These instances were selected, because they are among the largest instances an optimal solution is known for. The choice of instances was limited by the availability of optimal or at least high quality tours which were used to to evaluate the tours found in our experiments. Preferably, larger TSPLIB instances and benchmark instances from the DIMACS challenge [24] (E series instances) would have been included, too.

For each TSP instance 20 nearest-neighbor tours were constructed. Each of these tours was optimized to a local optimum by the local search heuristics 2-opt, 3-opt, Lin-Kernighan (LK-opt), and LK-Helsgaun (LK-H), respectively. For Helsgaun's LK parameter MAX_TRIALS was set to 100 (instead of number of cities). Totally, 600 tours were constructed to test the fixing heuristic when applied to tours of different quality levels.

Table 1. Probabilities (in percent) for edges in a tour to match certain criteria. The data is grouped by TSP instance and tour type. The eight left most columns contain the original and the conditional probabilites for the four edge fixation heuristics discussed in this paper.

Type		P(OPT)	P(MST)	P(OPT|MST)	P(NN)	P(OPT|NN)	P(<M)	P(OPT|<M)	P(CP)	P(OPT|CP)
brd14051	NN	59.12	70.44	73.22	58.29	74.82	54.64	71.07	40.11	71.68
	2-opt	65.13	73.58	75.78	60.39	77.33	53.28	74.67	38.49	75.70
	3-opt	70.49	76.20	78.47	62.73	79.70	52.17	77.77	37.14	78.83
	LK-opt	77.93	77.53	83.07	62.99	83.99	52.16	82.78	35.03	83.55
	LK-H	92.44	76.59	93.93	61.18	94.08	48.61	93.45	31.75	93.88
	optimal	100.00	75.33	100.00	59.43	100.00	47.83	100.00	30.34	100.00
d15112	NN	61.21	71.33	74.09	59.16	75.89	54.86	71.73	39.88	73.37
	2-opt	66.06	73.69	75.99	60.58	77.73	54.22	74.36	38.50	76.24
	3-opt	71.07	76.04	78.61	62.99	79.91	53.67	77.63	37.70	79.07
	LK-opt	77.50	77.01	82.63	63.12	83.71	52.38	82.03	35.83	83.09
	LK-H	92.11	76.15	93.51	61.33	93.83	50.65	93.05	32.59	93.67
	optimal	100.00	74.70	100.00	59.61	100.00	49.79	100.00	31.25	100.00
it16862	NN	61.35	72.20	74.44	60.06	76.04	52.60	70.59	32.56	73.59
	2-opt	66.72	74.66	76.64	61.90	78.10	53.82	74.34	30.56	76.29
	3-opt	71.87	76.94	79.33	64.34	80.37	54.63	79.06	29.08	79.34
	LK-opt	77.91	77.64	83.07	64.23	84.03	53.35	83.12	27.14	83.28
	LK-H	91.56	76.58	93.10	62.52	93.22	51.16	92.70	24.07	93.29
	optimal	100.00	75.07	100.00	60.59	100.00	49.92	100.00	22.89	100.00
d18512	NN	60.68	71.58	73.99	59.19	75.71	55.22	72.11	39.14	73.05
	2-opt	65.88	74.18	76.01	60.77	77.62	53.83	75.06	37.94	76.13
	3-opt	70.80	76.68	78.40	63.05	79.74	53.23	77.94	37.15	78.88
	LK-opt	77.95	77.76	82.95	63.15	83.97	52.94	82.73	35.05	83.38
	LK-H	92.70	76.66	94.08	61.16	94.22	50.03	93.67	31.72	94.01
	optimal	100.00	75.34	100.00	59.41	100.00	49.06	100.00	30.34	100.00
sw24978	NN	65.39	76.00	75.24	65.38	75.78	42.56	74.30	24.96	74.83
	2-opt	68.18	75.47	76.95	64.14	77.67	41.14	75.22	25.04	76.94
	3-opt	72.26	76.91	79.09	65.45	79.67	41.82	77.55	24.30	79.11
	LK-opt	77.22	77.00	82.38	64.46	83.08	41.19	80.97	23.18	82.45
	LK-H	90.33	76.01	92.06	62.90	92.13	40.31	90.66	20.54	92.56
	optimal	100.00	74.38	100.00	60.85	100.00	39.15	100.00	19.36	100.00

Each heuristics' selection scheme was applied to the set of 600 tours. Average values over each set of 20 tours with the same setup were taken and summarized in Tab. 1. Here, the first column sets the instance under consideration. The 'Type' column determines in which local optimum the tours are located. Column 'P(OPT)' shows for a local optimal tour the percentage of edges that also occur in the known global optimal tour. Columns 'P(MST)', 'P(NN)', 'P(<M)', and 'P(CP)' contain the probability that a edge from a local optimal tour matches the given property. Columns 'P(OPT|MST)', 'P(OPT|NN)', 'P(OPT|<M)', and 'P(OPT|CP)' contain conditional probabilities, that an edge in a local optimal tour is part of the global optimal tour, given that it matches the properties MST, NN, <M, or CP, respectively.

Column 'P(OPT)' shows the percentage of edges from a local optimal tour occurring in the global optimal tour. The better a tour construction and improvement heuristics works, the more edges the resulting tour has in common with the optimal tour. Whereas tours constructed by a nearest neighbor construction heuristics share about 60–65% of all edges with optimal tours (depending on the instance), tours found by subsequently applied local search algorithms are better, ranging from 65–70% for 2-opt to more than 90% for LK-H-opt tours.

As each edge selection strategy has different criteria how to select edges, they differ in the number of optimal edges. The most edges get chosen by the MST strategy (column 'P(MST)') selecting about 70–80% of all tour edges. Other strategies select less edges: From NN with 60–65% down to <M and CP with 45–55% and 25–40% of all edges, respectively. Interestingly, the number of selected edges decreases for all instances and all selection strategies when applied to high quality tours such as 'LK-H' and optimal tours.

However, the quantity (number of edges selected for fixing) is not the only criterion to rate a selection strategy. Selected edges were checked whether they occur in the corresponding known optimal tour, too. When applying a fixing heuristics to a sub-optimal tour, a good heuristics should more likely select edges that occur in the optimal tour, too, rather than edges that do not occur in the optimal tour. Therefore we were interested in the probability that a sub-optimal tour's edge selected by a edge selection strategy would actually be contained in an optimal tour ('true positive') rather than being a 'false positive'.

Edge selection strategies tend to be more successful for better tours. For every selection strategy, the percentage of correctly selected edges (edges that occur in the optimal tour, too) increases with the tour quality. E. g. for nearest neighbor tours of instance it16862, only about 74.4% of all selected edges are optimal tour edges, too, but for LK-H optimal tours, the percentage of selected edges is much higher (93.1%). Comparing selection strategies, the nearest neighbor selection strategy (NN) has the best probability values for all combinations except for three cases, where the close pairs strategy (CP) is more likely to find the right edges for LK-H tours. Especially for lower quality tours (NN and 2-opt), selection strategies <M and CP have the lowest conditional probabilities, but this effect disappears with higher quality tours.

4   Experimental Setup and Results

For the experimental evaluation, the fixing heuristics have been integrated into a simple TSP solver written in Java. The solver works as follows: Each tour was reduced using one of the fixing heuristics and subsequently improved by an iterated local search (ILS) algorithm. In each iteration of the algorithm the current tour was perturbed and locally optimized by an LK implementation, which is able to handle fixed edges. For the perturbation a variable-strength *double-bridge move* (DBM) was used increasing the number of DBMs each two non-improving iterations. At the end of each iteration the new tour was compared to the previous tour and discarded if no improvement was found, otherwise kept for subsequent iterations. The iterations would stop after 2 non-improving iterations. Finally, the improved tour was expanded back to a solution for the original TSP instance. For comparison, all tours were optimized by the iterated local search algorithm without any reduction, too. This solver was not designed to compete with state-of-the-art solvers, but merely to evaluate our fixation heuristics. Each parameter setup was tested by applying it to the tours described in Sec. 3; average values were used for the following discussion. Computation times are utilized CPU time on a 2.8 GHz Pentium 4 system with 1 GB memory running Linux.

Table 2 holds the results for the different setups applied to the ILS and is structured as follows: Rows are grouped by instance (brd14051 to sw24978) and by starting tour for the ILS ('NN' to 'LK-H'). The instances are ordered by number of cities, the starting tours are ordered by descending tour length. Columns are grouped into blocks, where the first block summarizes an ILS setup without any fixation and the subsequent blocks summarize ILS setups with each fixation heuristics ('MST' to 'Close Pairs'). Each column block consists of four columns: Improvement found when applying the ILS, required CPU time until termination, size of the reduced instance (normalized number of cities), and fraction of edges that are fixed (tabu for any ILS operation).

For every instance and each fixing heuristics (including no fixing) holds that the better the starting tour is, the smaller the improvements found by the ILS are. Applying our TSP solver to nearest neighbor tours (Tab. 2, rows with start tour type 'NN') results in improvements of more than 20% for most setups (columns 'Impr. [%]'). For better starting tours, less improvement is achieved, down to improvements of 0% for starting tours coming from LK-H.

Each fixing ILS setup can be compared with the corresponding ILS setup without fixing regarding improvement on the given start tour and the required CPU time. The following observations can be drawn from Tab. 2:

- For non-fixing setups, the CPU time is always higher compared to fixing setups, as the effective problem size is larger for the non-fixing setup. However, time consumption does not directly map to better tour quality.
- The Close Pairs (CP) fixing heuristics yields in improvements as good as for non-fixing ILS, but requires significantly less time to reach these quality levels. E. g. for instance brd14051 starting with 3-opt tours, both the non-fixing ILS and the CP fixing ILS improve the start tour by about 6.2%, but the CP-ILS requires only 31.9 s, whereas the non-fixing ILS requires 44.8 s.
- For the other fixing heuristics hold that they consume both less CPU time and result in lesser improvements. Although this makes comparing the different fixing strategies hard, improvements are still competitive while requiring significantly less CPU time compared to the non-fixing ILS.
- Among all fixation-based ILS setups, the MST heuristics results in both the smallest improvements and lowest running times compared to the other fixation heuristics. E. g. for instance sw2978 starting with 2-opt tours, the MST heuristics results in an improvement of 10.1% consuming 21.8 s, whereas all other fixation and non-fixation ILS setups find improvements of 11.4% and better consuming 41.0 s and more.
- Comparing CPU time consumption versus possible improvement, the fixation heuristics can be put into three groups: Between the expensive, but

Table 2. Results for different setups of instance, start tour, and fixing heuristics. The data is grouped by TSP instance and tour type. The result columns can be grouped into five blocks (for different fixing heuristics) consisting of four columns each: Achieved improvement from starting tour, utilized CPU time, fraction of free edges after fixing and size of the reduced instance (based on number of nodes).

Instance	Start Tour Type	No Fixing				MST				Nearest Neighbor				Light Median <				Close Pairs			
		Impr. [%]	CPU Time [s]	Free Edges [%]	Red. Size [%]	Impr. [%]	CPU Time [s]	Free Edges [%]	Red. Size [%]	Impr. [%]	CPU Time [s]	Free Edges [%]	Red. Size [%]	Impr. [%]	CPU Time [s]	Free Edges [%]	Red. Size [%]	Impr. [%]	CPU Time [s]	Free Edges [%]	Red. Size [%]
brd14051	NN	24.1	36.3	—	—	21.7	8.5	56.8	47.1	23.3	14.5	55.5	72.5	22.6	18.4	73.7	68.9	24.0	31.6	68.8	100.0
	2-opt	13.2	33.8	—	—	10.6	7.3	55.8	46.8	12.3	13.3	55.0	72.2	11.9	15.2	73.9	68.8	13.0	25.2	68.8	100.0
	3-opt	6.3	44.8	—	—	4.1	8.1	55.1	46.8	5.5	15.7	54.7	72.1	5.4	18.9	74.5	69.2	6.2	31.9	68.8	100.0
	LK-opt	0.1	19.6	—	—	0.0	4.0	55.0	49.0	0.0	7.7	56.0	75.5	0.1	8.8	74.4	68.1	0.1	13.8	69.2	100.0
	LK-H	0.0	11.4	—	—	0.0	2.9	54.4	45.6	0.0	5.5	55.1	73.5	0.0	6.4	75.6	69.1	0.0	9.4	69.6	100.0
d15112	NN	25.5	43.7	—	—	22.9	9.9	57.1	46.8	24.5	17.8	55.5	71.6	23.9	20.8	75.4	66.8	25.2	37.7	68.2	100.0
	2-opt	14.6	38.7	—	—	12.1	8.1	56.0	46.9	13.7	15.1	55.2	71.6	13.3	16.8	75.5	66.9	14.4	30.8	68.3	100.0
	3-opt	6.7	44.2	—	—	4.6	7.8	55.4	47.4	5.9	15.3	54.9	71.7	5.7	17.2	75.7	66.7	6.5	32.6	68.2	100.0
	LK-opt	0.1	22.1	—	—	0.0	4.3	55.2	49.6	0.0	8.2	56.2	75.3	0.0	8.7	75.8	66.5	0.1	15.1	68.6	100.0
	LK-H	0.0	12.4	—	—	0.0	3.2	54.3	46.6	0.0	5.8	55.2	73.5	0.0	6.5	76.4	65.9	0.0	10.0	68.9	100.0
fl16862	NN	24.2	51.1	—	—	21.7	10.5	56.4	45.6	23.4	17.7	55.2	70.2	23.6	23.3	71.4	70.7	24.3	41.4	75.8	100.0
	2-opt	13.5	60.6	—	—	11.2	11.2	55.5	45.5	12.7	20.4	54.9	70.1	12.9	24.7	71.7	70.7	13.5	46.7	75.9	100.0
	3-opt	6.5	57.0	—	—	4.5	9.2	54.9	46.0	5.8	17.7	54.5	70.1	6.1	22.0	72.0	71.2	6.4	41.7	76.1	100.0
	LK-opt	0.8	35.1	—	—	0.4	5.6	55.0	48.5	0.7	11.8	55.7	73.3	0.9	14.6	71.6	70.2	0.7	25.9	76.5	100.0
	LK-H	0.0	13.6	—	—	0.0	3.3	54.3	46.0	0.0	6.1	55.0	71.9	0.0	7.6	71.5	70.2	0.0	11.3	77.2	100.0
d18512	NN	24.2	37.5	—	—	21.7	8.9	56.7	46.6	23.4	15.3	55.3	72.4	22.8	19.9	72.8	70.2	24.1	32.8	68.9	100.0
	2-opt	13.7	49.8	—	—	11.2	10.6	55.8	46.7	12.8	19.5	55.0	72.4	12.6	22.9	73.0	70.1	13.6	39.4	68.9	100.0
	3-opt	6.8	47.6	—	—	4.6	8.9	55.1	47.0	5.9	18.0	54.7	72.4	5.9	19.8	73.1	69.8	6.6	34.6	68.8	100.0
	LK-opt	0.1	25.0	—	—	0.0	5.4	55.0	48.9	0.0	10.6	56.1	75.7	0.0	11.6	73.0	69.2	0.0	18.4	69.4	100.0
	LK-H	0.0	14.4	—	—	0.0	3.7	54.3	45.4	0.0	7.1	55.1	73.6	0.0	8.0	73.5	69.3	0.0	11.8	69.7	100.0
sw24978	NN	20.8	100.5	—	—	17.8	16.1	56.7	43.1	19.6	29.7	54.9	66.0	20.1	48.6	77.4	77.4	20.7	89.0	79.7	100.0
	2-opt	12.6	144.3	—	—	10.1	21.8	56.0	45.2	11.7	41.0	54.9	67.4	12.1	65.2	77.2	78.3	12.6	119.1	79.6	100.0
	3-opt	6.0	109.5	—	—	4.1	14.7	55.2	46.1	5.2	29.9	54.8	68.1	5.6	48.2	77.0	78.4	5.9	88.5	79.6	100.0
	LK-opt	0.5	57.2	—	—	0.1	8.2	55.3	49.4	0.2	16.3	56.1	72.2	0.3	26.7	77.1	79.9	0.3	46.2	80.0	100.0
	LK-H	0.0	17.7	—	—	0.0	4.9	54.7	47.2	0.0	8.3	55.3	70.8	0.0	12.7	76.9	79.2	0.0	16.9	80.7	100.0

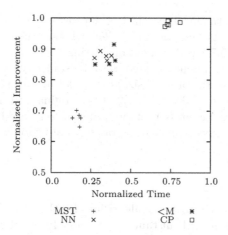

Fig. 3. For each of the five instances using 3-opt starting tours, improvements and CPU time of the ILS with fixation (MST, NN, <M, and CP) are normalized with the results from ILS runs without fixation

good CP heuristics and the cheap, but not so good MST heuristics the remaining two heuristics NN and <M can be located. These two "medium" heuristics show similar results both for found tour improvement and CPU time consumption.

The last observation can be visualized as in Fig. 3, which compares the improvements for each fixing heuristics applied to 3-opt tours. Both time and improvement are normalized for each of the five TSP instances by the values from the runs without fixing. The CP and the MST heuristics' values are separated from the central cluster consisting of NN and <M results. As can be seen, the CP heuristics reaches improvements as good as the non-fixing ILS (normalized improvement close to 1.0), but requires only $\frac{3}{4}$ of the time. NN and <M heuristics find improvements of about 90% of those from the non-fixing ILS and still demand less than the half of the time. The MST heuristics reaches about 70% of the non-fixing ILS's improvement, while consuming only a quarter of the CPU time.

For all fixing heuristics the size of the original instance has been compared to the corresponding reduced instances' size (in percent, columns 'Red. Size [%]' in Tab. 2) and the number of free edges within the reduced instance (in percent, columns 'Free Edges [%]').

– For close pairs (CP) fixations holds that the reduced instance's size equals always with the original instance's size, as fixed edges can not have neighboring edges that are fixed, too, as this would contradict the selection criterion. Thus, no nodes are redundant.
– For all combinations of instance and start tour, the MST fixing heuristics is the most progressive one resulting in the smallest instances. Here, fixed instances have on average less than half the number of cities compared to the original instances. Within these reduced instances, more than half of the

edges are free for the local search. E. g. for instance d15112 and 3-opt tours, only 47.4% of the original nodes are left, whereas the other heuristics leave 54.9% (NN) to 75.7% (<M) (not considering CP).

- The nearest neighbor heuristics reduces all instance types for about the same level (to 70–75% of the original size) except for instance sw24978, where the fixed instance reach 66.0% to 72.2% of the original instances' sizes.
- Over all instances and start tour type the nearest neighbor heuristics has a very stable percentage of free edges, ranging only between 54.2% and 56.2%.

5   Conclusion

In order to extend current memetic algorithms for the TSP to find close to optimum solutions for large TSP instances, we studied several edge-based problem reduction techniques that can easily be incorporated into an evolutionary local search framework. We have shown that fixing edges in TSP tours can considerably reduce the computation time of a TSP solver compared to applying the same solver to the unmodified problem instance. Still, the solutions found when using fixing heuristics are nearly as good as the solutions found without fixing. Therefore, edge fixing is a feasible approach to solve tours that are otherwise too large for solvers regarding memory or time consumption.

When selecting one of the proposed fixing heuristics, a trade-off between expected solution quality, computation time, or required preprocessing steps has to be made. E. g. the MST heuristics is expected to consume the least time, but requires building an MST in advance. The close pairs strategy can be used if no global knowledge is available, but here too few edges get fixed to decrease an instance's size considerably. As a compromise regarding time and quality, either the nearest neighbor or the lighter than median heuristics can be applied.

Future work will focus on developing fixing heuristics for population-based EAs and for very large TSP instances. For EAs, the knowledge of which edges occur in parent tours can be used to select edges in offspring tours. For very large instances such as the World TSP, fixation heuristics may exploit geographical properties.

References

1. Garey, M.R., Johnson, D.S.: Computers and Intractability: A Guide to the Theory of NP-Completeness. W. H. Freeman, San Francisco, CA, USA (1979)
2. Applegate, D., Bixby, R., Chvátal, V., Cook, W.: On the Solution of Traveling Salesman Problems. Documenta Mathematica **Extra Volume ICM III** (1998) 645–656
3. Applegate, D., Bixby, R., Chvátal, V., Cook, W.: Implementing the Dantzig-Fulkerson-Johnson Algorithm for large Traveling Salesman Problems. Mathematical Programming **97** (2003) 91–153
4. Lin, S., Kernighan, B.W.: An Effective Heuristic Algorithm for the Traveling-Salesman Problem. Operations Research **21**(2) (1973) 498–516

5. Johnson, D.S.: Local optimization and the traveling salesman problem. In: Proceedings of the 17th International Colloquium on Automata, Languages and Programming. Number 443 in Lecture Notes in Computer Science, Warwick University, England, Springer-Verlag (1990) 446–461
6. Moscato, P., Norman, M.G.: A Memetic Approach for the Traveling Salesman Problem Implementation of a Computational Ecology for Combinatorial Optimization on Message-Passing Systems. In Valero, M., Onate, E., Jane, M., Larriba, J.L., Suarez, B., eds.: Parallel Computing and Transputer Applications, Amsterdam, IOS Press (1992) 177–186
7. Merz, P., Freisleben, B.: Memetic Algorithms for the Traveling Salesman Problem. Complex Systems **13**(4) (2001) 297–345
8. Applegate, D., Bixby, R., Chvátal, V., Cook, W.: Concorde TSP Solver. urlhttp://www.tsp.gatech.edu/concorde/ (2005)
9. Cook, W., Seymour, P.: Tour Merging via Branch-Decomposition. INFORMS Journal on Computing **15**(3) (2003) 233–248
10. Applegate, D., Bixby, R., Chvátal, V., Cook, W.: Finding Cuts in the TSP (a Preliminary Report). Technical Report 95-05, Rutgers University, Piscataway NJ (1995)
11. Helsgaun, K.: An Effective Implementation of the Lin-Kernighan Traveling Salesman Heuristic. European Journal of Operational Research **126**(1) (2000) 106–130
12. Cook, W.J.: Log of SW24978 Computation. http://www.tsp.gatech.edu/world/swlog.html (2004)
13. Arenas, M.G., Collet, P., Eiben, A.E., Jelasity, M., Merelo, J.J., Paechter, B., Preuß, M., Schoenauer, M.: A Framework for Distributed Evolutionary Algorithms. In et al., J.J.M.G., ed.: Proceedings of the 7th International Conference on Parallel Problem Solving from Nature, PPSN VII. Volume 2439 of Lecture Notes in Computer Science., Springer, Berlin, Heidelberg (2002) 665–675
14. Fischer, T., Merz, P.: Embedding a Chained Lin-Kernighan Algorithm into a Distributed Algorithm. In: MIC'2005 – 6th Metaheuristics International Conference, Vienna, Austria (2005)
15. Cook, W.J.: World Traveling Salesman Problem. http://www.tsp.gatech.edu/world/ (2005)
16. Walshaw, C.: Multilevel Refinement for Combinatorial Optimisation Problems. Annals of Operations Research **131** (2004) 325–372
17. Walshaw, C.: A Multilevel Approach to the Travelling Salesman Problem. Operations Research **50**(5) (2002) 862–877
18. Applegate, D., Cook, W., Rohe, A.: Chained Lin-Kernighan for Large Traveling Salesman Problems. INFORMS Journal on Computing **15**(1) (2003) 82–92
19. Kruskal, J.B.: On the Shortest Spanning Subtree of a Graph and the Traveling Salesman Problem. Proceedings of the American Mathematical Society **7** (1956) 48–50
20. Friedman, J.H., Baskett, F., Shustek, L.H.: An Algorithm for Finding Nearest Neighbors. IEEE Transactions on Computers (TOC) **C-24** (1975) 1000–1006
21. Sproull, R.F.: Refinements to Nearest-Neighbor Searching in k-Dimensional Trees. Algorithmica **6**(4) (1991) 579–589
22. Reinelt, G.: TSPLIB — a traveling salesman problem library. ORSA Journal on Computing **3**(4) (1991) 376–384 See also http://www.iwr.uni-heidelberg.de/groups/comopt/software/TSPLIB95/.
23. Cook, W.J.: National Traveling Salesman Problems. http://www.tsp.gatech.edu/world/countries.html (2005)
24. Johnson, D.S., McGeoch, L.A. In: Experimental Analysis of Heuristics for the STSP. Kluwer Academic Publishers (2002) 369–443

Iterated k-Opt Local Search for the Maximum Clique Problem

Kengo Katayama, Masashi Sadamatsu, and Hiroyuki Narihisa

Information and Computer Engineering,
Okayama University of Science
1 - 1 Ridai-cho, Okayama, 700-0005 Japan
{katayama, masashi, narihisa}@ice.ous.ac.jp

Abstract. This paper presents a simple iterated local search meta-heuristic incorporating a k-opt local search (KLS), called *Iterated KLS* (*IKLS* for short), for solving the maximum clique problem (MCP). IKLS consists of three components: LocalSearch at which KLS is used, a Kick called LEC-Kick that escapes from local optima, and Restart that occasionally diversifies the search by moving to other points in the search space. IKLS is evaluated on DIMACS benchmark graphs. The results showed that IKLS is an effective algorithm for the MCP through comparisons with multi-start KLS and state-of-the-art metaheuristics.

1 Introduction

Let $G = (V, E)$ be an arbitrary undirected graph where V is the set of n vertices and $E \subseteq V \times V$ is the set of edges in G. For a subset $S \subseteq V$, let $G(S) = (S, E \cap S \times S)$ be *the subgraph* induced by S. A graph $G = (V, E)$ is *complete* if all its vertices are pairwise adjacent, i.e., $\forall\, i, j \in V$ with $i \neq j$, $(i, j) \in E$. A *clique* C is a subset of V such that the induced graph $G(C)$ is complete. The objective of the maximum clique problem (MCP) is to find a clique of maximum cardinality in G. See the recent survey [3] on the MCP, which also contains an extensive bibliography.

The MCP is known to be $\mathcal{NP}$-hard [4] for arbitrary graphs, and strong negative results have been shown. Håstad [6] showed that if $\mathcal{NP} \neq \mathcal{ZPP}$ then no polynomial time algorithm can approximate the maximum clique within a factor of $n^{1-\epsilon}$ for any $\epsilon > 0$, and this margin was tightened by Khot [13] to $n/2^{(\log n)^{(1-\epsilon)}}$.

Several exact methods such as branch-and-bound algorithms [23,20] have been proposed to solve the MCP exactly. However, their effectiveness and applicability are limited to relatively small or sparse graphs. Therefore much effort has been directed towards devising efficient heuristic and metaheuristic algorithms to find near-optimal solutions to large (dense) graphs within reasonable times. A collection of them can be found in [8,3]. More recently, the following promising metaheuristic algorithms have been proposed: Reactive Local Search [2], Variable Neighborhood Search [5], Steady-State Genetic Algorithm [22], and Dynamic Local Search [21].

C. Cotta and J. van Hemert (Eds.): EvoCOP 2007, LNCS 4446, pp. 84–95, 2007.

Most of the (meta)heuristic approaches to the MCP are based on a basic heuristical principle that prefers vertices of higher degrees to vertices of lower degrees in order to find a larger clique. Heuristics based on the principle are called *greedy*. The greedy heuristic can be regarded as a local search (or local improvement) method. The basic procedure is as follows: Given a current clique CC having a single vertex $v \in V$, one of the vertices in vertex set PA of the highest degree given by $deg_{G(PA)}$ is repeatedly added to expand CC until PA is empty, where PA denotes the vertex set of possible additions, i.e., the vertices that connected to all vertices of CC, and $deg_{G(S)}(v)$ stands for the degree of a vertex $v \in S$ in the subgraph $G(S)$, where $S \subseteq V$. This greedy method was called *1-opt local search* in [9] because at each iteration t, a single vertex $v \in PA^{(t)}$ is moved to $CC^{(t)}$ to obtain a larger clique $CC^{(t+1)} := CC^{(t)} \cup \{v\}$.

The 1-opt local search has been generalized in [9] by moving multiple vertices, instead of a single vertex, at each iteration, in order to obtain a better clique. The generalized local search is called *k-opt local search* (*KLS*). In KLS, variable, not fixed, k vertices are moved to or from a current clique simultaneously at each iteration by applying a sequence of add and drop move operations. The idea of the sequential moves in KLS is borrowed from *variable depth search* (VDS) of Lin and Kernighan [12,14]. KLS has been tested on the 37 hard DIMACS benchmark graphs [8] and compared with recent metaheuristics of reactive local search [2], genetic local search and iterated local search [16]. The results showed that KLS was capable of finding better or at least competitive cliques in reasonable times.

Since KLS can be regarded as a simple local improvement tool and there is no parameter setting by user, such as one to compulsorily stop KLS [9], it can be used, without serious modification for the algorithm, as a component in metaheuristic frameworks such as iterated local search [15], memetic algorithm [19], etc. Similar concepts can be found in several metaheuristics incorporating Lin-Kernighan heuristic (VDS based local search) for the traveling salesman problem (TSP): iterated Lin-Kernighan [7], chained Lin-Kernighan [1], and genetic iterated local search [10]. For other hard problems, several metaheuristics with VDS based local search have been proposed for the graph partitioning problem [17] and for the unconstrained binary quadratic programming problem [11,18]. These algorithms are known to be one of the best available metaheuristic algorithms to the TSP and other hard problems. It therefore seems that their high search performances mainly depend on effectiveness of the local search part of which the algorithms are composed. Judging from these contributions and dependability, the applications of metaheuristics embedded with KLS to the MCP appear to be effective and promising.

In this paper we propose an iterated local search incorporating KLS, called *Iterated KLS* (*IKLS* for short) for the MCP. IKLS consists of three components: LOCALSEARCH KLS, KICK called *Lowest-Edges-Connectivity-based Kick* (*LEC-Kick* for short), and RESTART; the details can be found in Section **2**. We evaluate the performance of IKLS on the DIMACS graphs, and show that despite the simplicity of the procedure, IKLS is effective through comparisons with multistart KLS and the promising metaheuristics described above for the MCP.

```
      procedure IKLS
      input: graph G = (V, E);
      output: best clique C_best in G;
      begin
1         generate C; compute PA, OM, and deg_G(PA);
2         C := KLS(C, PA, OM, deg_G(PA)); C_best := C;
3         repeat
4             C := LEC-Kick(C, PA, OM, deg_G(PA));
5             C := KLS(C, PA, OM, deg_G(PA));
6             if |C| > |C_best| then C_best := C; endif
7             if restart=true then
8                 generate C; compute PA, OM, and deg_G(PA);
9                 C := KLS(C, PA, OM, deg_G(PA));
10                if |C| > |C_best| then C_best := C; endif
11            endif
12        until terminate=true;
13        return C_best;
      end;
```

Fig. 1. The flow of Iterated KLS

2   Iterated k-Opt Local Search

Iterated local search [15] can be thought of as a simple and powerful metaheuristic that repeatedly applies local search technique to solutions which are obtained by kick technique (corresponding to a mutation) that escapes from previously found local optima. In this section we show an iterated local search incorporating KLS, called *Iterated k-opt Local Search* (*IKLS* for short), for the MCP.

2.1   IKLS

Given a local optimum (clique), each iteration of IKLS consists of LocalSearch at which KLS is used and Kick that escapes from local optima obtained by KLS. As the additional strategy performed occasionally, we use Restart that diversifies the search by moving to other positions in the search space. The top level flow of IKLS is shown in Figure 1.

At first (line 1), we generate a feasible solution (clique) C that contains a single vertex selected from V at random. We then compute the associated PA, OM, and $deg_{G(PA)}$[1], where OM denotes, given a current clique CC, the vertex set of one edge missing, i.e., the vertices that are connected to $|CC| - 1$ vertices of CC, provided that $CC \subseteq OM$. KLS is applied to C, and the resulting local optimum is stored as the best clique C_{best}. The loop (lines 3–12) of IKLS that includes the three components is repeated until the stopping condition (line 12) is satisfied. In our experiments, we stop IKLS when the clique with target size that is optimal (or best-known) for each graph is found or the maximum execution number of local searches is equal to $100 \times n$, where n is the number of vertices of G.

The details of LocalSearch, Kick, and Restart components are described in the following subsections.

---

[1] The information of PA, OM, and $deg_{G(PA)}$ is updated whenever a single vertex is moved to or from current clique during IKLS. The updating technique and data structure used in IKLS are derived from the literature [2].

```
      KLS(CC, PA, OM, deg_G(PA))
      begin
1        repeat
2           CC_prev := CC, D := CC_prev, g := 0, gmax := 0;
3           P := {1, . . . , n}; // Note that some vertices are removed only for the first iteration; see 2.3.
4           repeat
5              if |PA ∩ P| > 0 then          // Add Phase
6                 find a vertex v with max_{v∈{PA∩P}} { deg_G(PA∩P)(v) };
7                 if multiple vertices with the same maximum degree are found
                  then select one vertex v among them randomly;
8                 CC := CC ∪ {v}, g := g + 1, P := P\{v};
9                 if g > gmax then gmax := g, CC_best := CC;
10             else          // Drop Phase (if {PA ∩ P} = ∅)
11                find a vertex v ∈ {CC ∩ P} such that the resulting |PA ∩ P| is maximized;
12                if multiple vertices with the same size of the resulting |PA ∩ P| are found,
                  then select one vertex v among them randomly;
13                CC := CC\{v}, g := g - 1, P := P\{v};
14                if v is contained in CC_prev then D := D\{v};
15             endif
16             update PA, OM, and deg_G(PA);
17          until D = ∅;
18          if gmax > 0 then CC := CC_best else CC := CC_prev;
19       until gmax ≤ 0;
20       return CC;
      end;
```

Fig. 2. The pseudo code of KLS performed in IKLS

2.2 Local Search

The work of LOCALSEARCH process in IKLS is to find local optima in a given graph G. In the process we use KLS at lines 2, 5 and 9 in Figure 1. In the following, we briefly review KLS [9], and describe simple devices given for KLS.

KLS performs several add and drop moves for a given feasible clique CC at each iteration. The add move is to add a vertex $v \in PA$ to CC if $PA \neq \emptyset$ and the drop move is to drop a vertex $v \in CC$. A set of neighbor cliques obtained by each of the 1-opt moves is called *1-opt neighborhood*.

The 1-opt neighborhood has been based in most of the existing (meta)heuristic algorithms [8,3,16,2,21] for the MCP. Therefore, it is natural to consider lager neighborhoods than the 1-opt one. However, it is confronted with several drawbacks, such as how many vertices should be moved at each iteration, because the feasible cliques and reasonable search are usually desired in local search; see [9] on details of the drawbacks. In [9], the drawbacks were removed by introducing *variable depth search* (VDS) [12,14] that is to change the size of the neighborhood adaptively so that the algorithm can effectively traverse larger search space within reasonable time. KLS determines dynamically at each iteration the value of k, where k is the number of vertices to move. KLS efficiently explores the *(variable) k-opt neighborhood* defined as the set of neighbors that can be obtained by a sequence of several add and drop moves that are adaptively changed in the feasible search space. More details and the basic procedure of KLS can be found in [9].

To perform effective searches with IKLS, two simple devices are given for the original KLS of [9]. The pseudo code of KLS for which the devices are given is shown in Figure 2. Note that the devices lead to no serious modifications and the basic procedure is the same as the original KLS.

The first one is to alter the process on the vertex selection rule performed in the add and drop phases of KLS. In KLS used in IKLS, we select a vertex of

```
        LEC-Kick(CC, PA, OM, deg_G(PA))
     begin
1      if all i ∈ CC are disconnected to all j ∈ V\CC then
2          select a vertex v ∈ V\CC randomly; compute PA, OM, and deg_G(PA);
3          CC := ∅; CC := CC ∪ {v}; return new clique CC;
4      endif
5      find a vertex v ∈ V\CC with the lowest edge number to vertices of CC.
6      if multiple vertices with the same lowest edge number are found
       then select one vertex v among them randomly;
7      drop vertices from CC that are not connected to v;
       // the dropped vertices are removed from P in Fig. 2 (line 3) only for 1st iteration of the next KLS.
8      update PA, OM, and deg_G(PA);
9      return new clique CC;
     end;
```

Fig. 3. The pseudo code of Kick process in IKLS

the highest degree in subgraph $G(PA \cap P)$ at line 6 instead of $G(PA)$ in the add phase, and select a vertex from the current clique in the drop phase such that resulting $|PA \cap P|$ instead of $|PA|$ is maximized at line 11. This simple device in KLS contributes to slightly improve average size of cliques obtained by it without increasing the computation time in many cases.

The second device (line 3) is related to the following KICK process.

2.3 Kick

The role of KICK is to escape from local optima found by LOCALSEARCH process by moving to other points where are not so far from the local optima in the search space. This moving is made by perturbing the current local optimum slightly so that a different local optimum can be found by the forthcoming local search. Therefore, the general desire in designing KICK for a specific problem is to reduce the probability to fall back into a previously found local optimum without moving to a search point where is far away from the current one.

In the traveling salesman problem (TSP), a well-known kick method called *double-bridge move* [1,7,15] has been used for iterated local search and other relevant metaheuristic approaches in which a local search such as Lin-Kernighan heuristic [14] is incorporated. It is known to be very useful in that it avoids to fall back into a local optimum just found by the previous local search.

Since the double-bridge kick cannot be applied to the MCP, we newly design a simple kick method for IKLS. One of the simplest methods is, given a current clique CC, to drop m ($1 \leq m < |CC|$) vertices from CC. However, it is difficult to determine the number m that corresponds to *perturbation strength* [15] because a suitable and optimal m may depend on graph, clique given, etc. Therefore, the perturbation strength should be determined adaptively in IKLS. Our newly designed method, called *Lowest-Edges-Connectivity-based Kick* (*LEC-Kick* for short) is quite simple, and no parameter setting value such as m is required.

In IKLS, the KICK process (LEC-Kick) is performed at line 4 in Figure 1, and the detail is shown in Figure 3. Lines 1–4 show the exceptional processing performed only when all vertices of CC have no edge to all vertices of $V\backslash CC$ in G. In this case, we select a single vertex v from $V\backslash CC$ at random, and return the new (poor) clique having v.

Otherwise, the main process of LEC-Kick is performed at lines 5–9 in Figure 3. At first, we randomly select a vertex $v \in V \backslash CC$ with the *lowest edge* number to vertices of CC. After that, all vertices that are not connected to v are removed from CC to make a new clique at line 7. The new clique contains the vertex v that is not included in the initial clique given for the kick process. In this point, containing such a vertex v into a newly *kicked* clique to the forthcoming local search may contribute to reduce the probability to fall back into the clique just discovered by the previous local search.

Although the number of vertices dropped from CC is adaptively determined in the main LEC-Kick process, it may be feared that most of the vertices are dropped and therefore the resulting size of the clique kicked is too small. We will show the additional results (in Table 1) related to this concern in Section **3**.

To further reduce the probability described above, a simple device is given for the forthcoming KLS as the second device; the vertices dropped from CC at line 7 in Figure 3 are removed from the set P at line 3 in Figure 2. This device is performed only for the first iteration of the forthcoming KLS, not for all the iterations.

Let us concentrate our attention on KICK and LOCALSEARCH processes in IKLS. Both processes are sequentially performed in every IKLS iteration to repeatedly explore the cliques that exist around local optima. It therefore seems to be similar to the "plateau search" [2,21]. We here regard the sequential search performed by the two processes as *"sawteeth search"* instead of "plateau search" because the up-and-down motion imaged from fluctuations of the sizes of cliques searched by the sequential processes is relatively larger than that imaged from the sizes of cliques found in the *original* plateau search as in [2]. Note that restricted plateau searches are self-adaptively executed in KLS itself.

2.4 Restart

The aim of RESTART, the additional strategy performed occasionally in IKLS, is to diversify the main search of IKLS, i.e., the sawteeth search described at the previous subsection, by moving compulsorily to other points in the search space. Since such diversifications should be given after the sawteeth search was performed for a while, we occasionally perform RESTART process if the following condition is satisfied at line 7 in Figure 1: if no new best clique CC_{best} is found for more than $|CC_{best}|$ iterations of the sawteeth search in IKLS, where $|CC_{best}|$ is the size of the best clique CC_{best} found so far in the search. In response to this requirement, a new clique is generated at line 8 by selecting a single vertex $v \in V \backslash \{CC_{best}\}$ at random. At line 9, the new clique C, the single vertex selected, is expanded by KLS, and the sawteeth search is started again.

3 Experimental Results

In order to evaluate the performance of IKLS, we performed extensive computational experiments on the 37 hard instances of DIMACS benchmark graphs with up to 4000 nodes and up to 5506380 edges, available from the Second DI-MACS Implementation Challenge[2]. All experiments were performed on Hewlett-Packard xw4300 workstation with Pentium4 3.4GHz, 4GB RAM, and Fedora Core 5, using the gcc compiler 4.11 with '-O2' option. To execute the DIMACS Machine Benchmark[3], this machine required 0.45 CPU seconds for r300.5, 2.78 (s) for r400.5 and 8.32 (s) for r500.5.

We first compare the performance of IKLS with that of multi-start KLS (MKLS for short) because it is reasonable in that both algorithms are based on the same local search KLS, and it may be useful to see the performance difference between the algorithmic frameworks. MKLS is a simple procedure, in which repeatedly KLS is applied to newly generated cliques at random, and the best overall clique is kept and output as the result. Each MKLS iteration consists of two parts: generating an initial (poor) clique having a single vertex from V at random and KLS that is applied to the initial clique. In our experiments, this simple process is repeated until the clique with target size that is known to be optimal (or best-known) for each graph is found or $100 \times n$ iterations. The stopping condition for MKLS is the same with IKLS (see **2.1**). The remaining parameter setting that should be set for IKLS has been described at **2.4**.

Table 1 shows the results of IKLS and MKLS. Each approach was run independently for 100 times on each graph, provided MKLS was carried out 10 runs instead of 100 runs only for MANN_a81 because of a large amount of computation times. The first two columns of the table contain the instance names and their best-known clique size "BR" ('*' if optimality is proved), respectively. In the following columns for IKLS results, we show the best found clique size "Best" with the number of times in which the best found cliques could be found by the algorithm "(#B)", the average clique size "Avg" with standard deviation "(s.dev.)", the worst clique size "Worst" with the number of times in which the worst cliques were found by the algorithm "(#W)", the average running time "Time(s)"[4] with standard deviation "(s.dev.)" in seconds in case the algorithm could find the best found cliques. The following three columns show the additional information in IKLS executions: the average steps of add moves "#add", the average number of kicks applied in a single IKLS run "#kick", and the average number of vertices dropped from current cliques in the kick process "#dk". In the remaining columns, we provide for MKLS results with the corresponding meaning which can be found in the columns of the IKLS results.

We observed in Table 1 that IKLS is capable of finding the best-known clique on all graphs, while MKLS fails on brock400_2, brock800_2, brock800_4, and keller6. It was shown that the average results "Avg" of IKLS were at least equal to, or better than those of MKLS except for C2000.9. In particular, the results of IKLS on MANN_a81 and keller6, known to be the hardest instances in the DIMACS graph set, were considerably better. Therefore the IKLS framework for the MCP seems to be better than the multi-start framework with KLS in

---

[2] http://dimacs.rutgers.edu/Challenges/

[3] dmclique is available from ftp://dimacs.rutgers.edu/pub/dsj/clique

[4] In Table 1 (and 2) the average run times less than 0.001 seconds are shown as "$< \epsilon$".

Table 1. Results of Iterated KLS and Multistart KLS on the DIMACS benchmark graphs

Instance	BR	IKLS (iterated k-opt local search)							MKLS (multistart k-opt local search)		
		Best (#B)	Worst (#W)	Avg (s.dev.)	Time(s) (s.dev.)	#add	#kick	#dk	Best (#B)	Avg (s.dev.)	Time(s)
C125.9	34*	34 (100)	34 (100)	34.00 (0)	<ε (0.001)	197	2	7	34 (100)	34.00 (0)	<ε
C250.9	44*	44 (100)	44 (100)	44.00 (0)	0.009 (0.009)	1761	17	10	44 (100)	44.00 (0)	0.023
C500.9	57	57 (100)	57 (100)	57.00 (0)	2.146 (1.973)	272172	2078	12	57 (100)	57.00 (0)	3.629
C1000.9	68	68 (90)	67 (10)	67.90 (0.300)	48.322 (46.010)	3943276	24027	14	68 (81)	67.81 (0.392)	130.012
C2000.9	78	78 (1)	75 (4)	76.11 (0.444)	236.405 (214.121)	11503314	59046	16	78 (1)	76.55 (0.517)	686.926
DSJC500.5	13*	13 (100)	13 (100)	13.00 (0)	0.052 (0.045)	1406	63	9	13 (100)	13.00 (0)	0.018
DSJC1000.5	15*	15 (100)	15 (100)	15.00 (0)	2.361 (2.245)	21217	854	11	15 (100)	15.00 (0)	0.510
C2000.5	16	16 (100)	16 (100)	16.00 (0)	7.017 (7.521)	17703	646	12	16 (100)	16.00 (0)	1.725
C4000.5	18	18 (100)	18 (100)	18.00 (0)	2285.502 (2000.747)	1545476	51537	13	18 (96)	17.96 (0.195)	3839.559
MANN_a27	126*	126 (100)	126 (100)	126.00 (0)	0.029 (0.027)	4036	14	10	126 (100)	126.00 (0)	0.013
MANN_a45	345*	345 (81)	344 (19)	344.81 (0.392)	752.261 (659.438)	34252779	36028	15	345 (8)	344.08 (0.271)	71.985
MANN_a81	1100	1100 (100)	1100 (100)	1100.00 (0)	512.354 (498.422)	5336497	1698	26	1100 (2/10)	1099.20 (0.400)	11877.433
brock200.2	12*	12 (100)	12 (100)	12.00 (0)	0.081 (0.077)	7100	392	8	12 (3)	11.03 (0.170)	0.046
brock200.4	17*	17 (100)	17 (100)	17.00 (0)	0.169 (0.171)	30669	995	9	17 (3)	16.03 (0.170)	0.030
brock400.2	29*	29 (72)	25 (28)	27.88 (1.795)	4.006 (4.414)	569995	10750	11	25 (100)	25.00 (0)	0.169
brock400.4	33*	33 (100)	33 (100)	33.00 (0)	1.308 (1.144)	187337	3521	11	33 (18)	26.44 (3.073)	2.154
brock800.2	24	24 (5)	21 (95)	21.15 (0.653)	3.207 (7.924)	181434	4509	12	21 (100)	21.00 (0)	0.566
brock800.4	26	26 (11)	21 (89)	21.55 (1.564)	5.861 (12.259)	326326	8185	12	21 (100)	21.00 (0)	1.413
gen200.p0.9.44	44*	44 (100)	44 (100)	44.00 (0)	0.008 (0.006)	1967	21	9	44 (100)	44.00 (0)	0.024
gen200.p0.9.55	55*	55 (100)	55 (100)	55.00 (0)	0.003 (0.003)	652	6	9	55 (100)	55.00 (0)	0.004
gen400.p0.9.55	55	55 (100)	55 (100)	55.00 (0)	2.506 (2.552)	370474	3022	12	55 (76)	54.52 (0.854)	13.754
gen400.p0.9.65	65	65 (100)	65 (100)	65.00 (0)	0.012 (0.010)	1716	12	12	65 (100)	65.00 (0)	0.045
gen400.p0.9.75	75	75 (100)	75 (100)	75.00 (0)	0.010 (0.007)	1190	8	11	75 (100)	75.00 (0)	0.033
hamming8-4	16*	16 (100)	16 (100)	16.00 (0)	<ε (0.001)	27	8	0	16 (100)	16.00 (0)	<ε
hamming10-4	40	40 (100)	40 (100)	40.00 (0)	0.013 (0.009)	871	8	12	40 (100)	40.00 (0)	0.122
keller4	11*	11 (100)	11 (100)	11.00 (0)	<ε (0.001)	24	0	0	11 (100)	11.00 (0)	<ε
keller5	27	27 (100)	27 (100)	27.00 (0)	0.024 (0.031)	1528	35	11	27 (100)	27.00 (0)	0.023
keller6	59	59 (100)	59 (100)	59.00 (0)	8.055 (6.983)	178369	1844	18	57 (100)	57.00 (0)	830.578
p_hat300-1	8*	8 (100)	8 (100)	8.00 (0)	0.002 (0.002)	125	9	5	8 (100)	8.00 (0)	0.001
p_hat300-2	25*	25 (100)	25 (100)	25.00 (0)	0.001 (0.001)	112	1	17	25 (100)	25.00 (0)	<ε
p_hat300-3	36*	36 (100)	36 (100)	36.00 (0)	0.004 (0.003)	720	8	15	36 (100)	36.00 (0)	0.008
p_hat700-1	11*	11 (100)	11 (100)	11.00 (0)	0.056 (0.053)	1173	69	7	11 (100)	11.00 (0)	0.066
p_hat700-2	44*	44 (100)	44 (100)	44.00 (0)	0.005 (0.003)	358	2	36	44 (100)	44.00 (0)	0.004
p_hat700-3	62	62 (100)	62 (100)	62.00 (0)	0.009 (0.007)	713	4	25	62 (100)	62.00 (0)	0.007
p_hat1500-1	12*	12 (100)	12 (100)	12.00 (0)	11.110 (10.899)	73540	3752	8	12 (100)	12.00 (0)	1.387
p_hat1500-2	65	65 (100)	65 (100)	65.00 (0)	0.039 (0.034)	1547	8	41	65 (100)	65.00 (0)	0.031
p_hat1500-3	94	94 (100)	94 (100)	94.00 (0)	0.128 (0.092)	5700	24	35	94 (100)	94.00 (0)	0.295

Table 2. Results of IKLS, RLS, DLS, VNS, and HSSGA on the DIMACS benchmark graphs

Instance	BR	IKLS			RLS [2]			DLS [21]			VNS [5]			HSSGA [22]		
		Best	Avg	Time(s)	Best	Avg	Time(s)	Best	Avg	Time(s)	Best	Avg	Time(s)	Best	Avg	Time(s)
C125.9	34*	34	34.0	$<\epsilon$	34	34.0	$<\epsilon$	34	34.0	$<\epsilon$	34	34.0	0.001	34	34.0	0.017
C250.9	44*	44	44.0	0.009	44	44.0	0.004	44	44.0	$<\epsilon$	44	44.0	0.019	44	43.8	0.097
C500.9	57	57	57.0	2.146	57	57.0	0.526	57	57.0	0.072	57	57.0	0.317	56	54.2	1.190
C1000.9	68	68	67.9	48.322	68	68.0	7.014	68	68.0	2.551	68	68.0	6.223	66	64.1	4.073
C2000.9	78	78	76.1	236.405	78	77.5	138.638	78	77.9	111.038	78	77.2	27.321	74	71.0	33.587
DSJC500.5	13*	13	13.0	0.052	13	13.0	0.032	13	13.0	0.007	13	13.0	0.078	13	13.0	0.202
DSJC1000.5	15*	15	15.0	2.361	15	15.0	1.086	15	15.0	0.459	15	15.0	1.278	15	14.7	2.106
C2000.5	16	16	16.0	7.017	16	16.0	1.679	16	16.0	0.556	16	16.0	1.683	16	15.4	7.855
C4000.5	18	18	18.0	2285.502	18	18.0	367.593	18	18.0	104.147	18	18.0	373.317	17	16.8	45.223
MANN_a27	126*	126	126.0	0.029	126	126.0	0.524	126	126.0	0.027	126	126.0	0.006	126	125.5	0.904
MANN_a45	345*	345	344.8	752.261	345	343.6	67.145	344	344.0	29.859	345	344.5	1.823	343	342.6	18.626
MANN_a81	1100	1100	1100.0	512.354	1098	1098.0	476.659	1098	1097.9	151.716	1100	1099.3	78.661	1095	1094.2	1140.894
brock200_2	12*	12	12.0	0.081	12	12.0	1.617	12	12.0	0.013	12	11.3	0.092	12	12.0	0.082
brock200_4	17*	17	17.0	0.169	17	17.0	3.281	17	17.0	0.026	17	16.9	0.596	17	16.7	0.325
brock400_2	29*	29	27.8	4.006	29	26.0	7.087	29	29.0	0.274	29	27.4	5.015	29	25.1	0.670
brock400_4	33*	33	33.0	1.308	33	32.4	18.292	33	33.0	0.038	33	33.0	3.236	33	27.0	0.787
brock800_2	24	24	21.1	3.207	21	21.0	0.797	24	24.0	9.041	21	21.0	1.033	21	20.7	3.060
brock800_4	26	26	21.5	5.861	21	21.0	1.127	26	26.0	5.103	21	21.0	3.795	21	20.1	0.867
gen200_p0.9_44	44*	44	44.0	0.008	44	44.0	0.006	44	44.0	$<\epsilon$	44	44.0	0.079	44	43.1	0.305
gen200_p0.9_55	55*	55	55.0	0.003	55	55.0	0.002	55	55.0	$<\epsilon$	55	55.0	0.021	55	55.0	0.082
gen400_p0.9_55	55	55	55.0	2.506	55	55.0	0.202	55	55.0	0.014	55	54.8	3.006	53	51.4	0.522
gen400_p0.9_65	65	65	65.0	0.012	65	65.0	0.008	65	65.0	$<\epsilon$	65	65.0	0.106	65	63.8	0.488
gen400_p0.9_75	75	75	75.0	0.010	75	75.0	0.008	75	75.0	$<\epsilon$	75	75.0	0.087	75	75.0	0.550
hamming8-4	16*	16	16.0	$<\epsilon$	16	16.0	$<\epsilon$	16	16.0	$<\epsilon$	16	16.0	$<\epsilon$	16	16.0	0.002
hamming10-4	40	40	40.0	0.013	40	40.0	0.013	40	40.0	0.004	40	40.0	0.022	40	39.0	2.914
keller4	11*	11	11.0	$<\epsilon$	11	11.0	$<\epsilon$	11	11.0	$<\epsilon$	11	11.0	$<\epsilon$	11	11.0	0.002
keller5	27	27	27.0	0.024	27	27.0	0.028	27	27.0	0.011	27	27.0	0.033	27	26.9	1.153
keller6	59	59	59.0	8.055	59	59.0	31.96	59	59.0	97.969	59	58.2	21.509	57	54.2	89.820
p_hat300-1	8*	8	8.0	0.002	8	8.0	0.003	8	8.0	$<\epsilon$	8	8.0	0.001	8	8.0	0.005
p_hat300-2	25*	25	25.0	0.001	25	25.0	0.001	25	25.0	$<\epsilon$	25	25.0	$<\epsilon$	25	25.0	0.005
p_hat300-3	36*	36	36.0	0.004	36	36.0	0.003	36	36.0	$<\epsilon$	36	36.0	0.006	36	35.9	0.051
p_hat700-1	11*	11	11.0	0.056	11	11.0	0.031	11	11.0	0.010	11	11.0	0.045	11	11.0	0.291
p_hat700-2	44*	44	44.0	0.005	44	44.0	0.004	44	44.0	$<\epsilon$	44	44.0	0.004	44	44.0	0.054
p_hat700-3	62	62	62.0	0.009	62	62.0	0.005	62	62.0	$<\epsilon$	62	62.0	0.005	62	61.7	0.573
p_hat1500-1	12*	12	12.0	11.110	12	12.0	5.097	12	12.0	1.555	12	12.0	36.454	12	11.5	4.173
p_hat1500-2	65	65	65.0	0.039	65	65.0	0.026	65	65.0	0.003	65	65.0	0.024	65	64.9	0.579
p_hat1500-3	94	94	94.0	0.128	94	94.0	0.032	94	94.0	0.005	94	94.0	0.050	94	93.1	0.830

terms of solution qualities although MKLS reaches the optimal or best-known clique in shorter times on several graphs, such as DSJC1000.5, p_hat1500-1, etc.

As shown in subsection **2.3**, although the number of vertices dropped from current clique in LEC-Kick is adaptively determined with the edge density of a given graph G and the connectivity of the current clique to a vertex v selected from $V \backslash CC$ at random, it is feared that most of the vertices are dropped in LEC-Kick and the size of the resulting clique is too small. The additional results of the column "#dk" in Table 1 wipe out the fear; for example, for MANN_a81 that is a dense graph and for DSJC500.5 that is a sparse (not dense) graph with edge density of 0.5, the average numbers of vertices dropped from current cliques are 26 and 9 (that correspond to 2.36% and 69.2% in each BR of the graphs), respectively. If we suppose that the optimal clique size depends on the edge density of G and in addition, KLS obtains near-optimum cliques in many iterations of IKLS, the perturbation strength in LEC-Kick (the results of "#dk") is quite reasonable in many cases to escape from local optima.

We next compare the performance of IKLS with those of state-of-the-art promising metaheuristics with which good results have been reported: Reactive Local Search (RLS) [2], Dynamic Local Search (DLS) [21], Variable Neighborhood Search (VNS) [5], and Steady-State Genetic Algorithm (HSSGA) [22]. Table 2 summarizes the results (Best, Avg, Time(s)) of IKLS, RLS, DLS, VNS and HSSGA. To have fair comparisons, the average running times of the competitors shown in the table were all adjusted according to the results based on the DIMACS Machine Benchmark test reported in their papers.

It is considerably impressive that the cliques of size 1100 on MANN_a81, known to be one of the hardest graphs in the benchmark set, can be obtained by IKLS in *all* 100 runs in relatively short running times, while RLS and DLS only finds the cliques of size 1098 as the best result in 100 runs, and VNS obtains the average size of 1099.3 in 10 runs with the clique sizes ranging from 1098 to 1100. Therefore, the capability of IKLS to find the best-known cliques of size 1100 for MANN_a81 is superior to the recent metaheuristics. Furthermore, it can be observed that IKLS finds the best-known cliques of size 59 on keller6 in all runs in shorter running times. From these results, it should be noted that IKLS is more efficient and effective than the others for such hard graphs.

The results of IKLS for the other graphs seem to be comparable with those of RLS, DLS, and VNS except for larger and denser C graphs such as C2000.9. For larger brock graphs such as brock800_2, brock800_4, etc., DLS clearly outperforms IKLS and the others on the average clique sizes obtained.

Although in this paper the results of IKLS have been shown only for the 37 graphs chosen from the 80 DIMACS benchmark ones, additional experimental results showed that IKLS obtained the best-known cliques in all 100 runs for the remaining graphs except for larger brock graphs. In addition, we have already observed that RESTART in IKLS contributes to improve the total performance on larger MANN_a graphs in comparison to IKLS without RESTART. Finally, in IKLS we adopted the random walk acceptance criterion [15], where a new solution found by KLS is perturbed by LEC-Kick in each IKLS iteration without taking into account the quality of solutions in the acceptance decision. Additional experimental results indicated that an alternative procedure of IKLS in which C_{best} (the best clique found so far) is perturbed by LEC-Kick was inferior to IKLS shown in the paper on C1000.9, C4000.5, MANN_a45, MANN_a81, gen400_p0.9_55, keller6, and all brock graphs particularly.

4 Conclusion

We proposed a simple iterated local search metaheuristic, called Iterated k-opt Local Search (IKLS), for solving the maximum clique problem (MCP). Each iteration of IKLS has the components of LOCALSEARCH at which KLS is used and a KICK called *Lowest-Edges-Connectivity-based Kick* that adaptively determines the perturbation strength. Both processes, called the *sawteeth search*, are repeated for a while, and RESTART is performed occasionally in order to diversify the search by compulsorily moving to other points in the search space. After the diversification, the sawteeth search is started again. Finally the best clique (or the target sized one) found by IKLS is output as the result.

To see the performance difference between algorithmic frameworks based on the same local search, KLS, we first compared Iterated KLS (IKLS) with multistart KLS (MKLS) on the 37 DIMACS benchmark graphs. The difference was observed on in particular hard graphs, and we showed that the IKLS framework is suitable for solving the MCP. Furthermore, the results of IKLS were compared with recent results of effective metaheuristics on the benchmark graphs. It was demonstrated that although IKLS fails on some graphs, it is capable of finding better or at least competitive cliques despite the simplicity of the procedure. In particular, it was impressive that IKLS found in all runs the cliques of size 1100 on MANN_a81, one of the hardest graphs in the benchmark set, and obtained the best-known cliques in all runs on keller6 in shorter running times than the other metaheuristics. We therefore conclude that IKLS is effective and a new promising metaheuristic for the MCP.

References

1. D. Applegate, W. Cook, and A. Rohe. Chained Lin-Kernighan for large traveling salesman problems. *INFORMS Journal on Computing*, Vol. 15, No. 1, pp. 82–92, 2003.
2. R. Battiti and M. Protasi. Reactive local search for the maximum clique problem. *Algorithmica*, Vol. 29, No. 4, pp. 610–637, 2001.
3. I.M. Bomze, M. Budinich, P.M. Pardalos, and M. Pelillo. The maximum clique problem. In D.-Z. Du and P.M. Pardalos, editors, *Handbook of Combinatorial Optimization (suppl. Vol. A)*, pp. 1–74. Kluwer, 1999.
4. M.R. Garey and D.S. Johnson. *Computers and Intractability: A Guide to the Theory of NP-Completeness*. Freeman, New York, 1979.
5. P. Hansen, N. Mladenović, and D. Urošević. Variable neighborhood search for the maximum clique. *Discrete Applied Mathematics*, Vol. 145, No. 1, pp. 117–125, 2004.

6. J. Håstad. Clique is hard to approximate within $n^{1-\epsilon}$. *Acta Mathematica*, Vol. 182, pp. 105–142, 1999.

7. D.S. Johnson and L.A. McGeoch. The traveling salesman problem: A case study. In E. Aarts and J.K. Lenstra, editors, *Local Search in Combinatorial Optimization*, pp. 215–310. John Wiley & Sons, 1997.

8. D.S. Johnson and M.A. Trick. *Cliques, Coloring, and Satisfiability*. Second DIMACS Implementation Challenge, DIMACS Series in Discrete Mathematics and Theoretical Computer Science. American Mathematical Society, 1996.

9. K. Katayama, A. Hamamoto, and H. Narihisa. An effective local search for the maximum clique problem. *Information Processing Letters*, Vol. 95, No. 5, pp. 503–511, 2005.

10. K. Katayama and H. Narihisa. Iterated local search approach using genetic transformation to the traveling salesman problem. In *Proc. of the Genetic and Evolutionary Computation Conference*, pp. 321–328, 1999.

11. K. Katayama, M. Tani, and H. Narihisa. Solving large binary quadratic programming problems by effective genetic local search algorithm. In *Proc. of the Genetic and Evolutionary Computation Conference*, pp. 643–650, 2000.

12. B.W. Kernighan and S. Lin. An efficient heuristic procedure for partitioning graphs. *Bell System Technical Journal*, Vol. 49, pp. 291–307, 1970.

13. S. Khot. Improved inapproximability results for maxclique, chromatic number and approximate graph coloring. In *Proceedings of the 42nd IEEE symposium on Foundations of Computer Science*, pp. 600–609, 2001.

14. S. Lin and B.W. Kernighan. An effective heuristic algorithm for the traveling salesman problem. *Operations Research*, Vol. 21, pp. 498–516, 1973.

15. H.R. Lourenço, O.C. Martin, and T. Stützle. Iterated local search. In F. Glover and G. Kochenberger, editors, *Handbook of Metaheuristics*, Vol. 57 of *International Series in Operations Research & Management Science*, pp. 321–353. Kluwer Academic Publishers, Norwell, MA, 2003.

16. E. Marchiori. Genetic, iterated and multistart local search for the maximum clique problem. In *Applications of Evolutionary Computing, LNCS 2279*, pp. 112–121. Springer-Verlag, 2002.

17. P. Merz and B. Freisleben. Fitness landscapes, memetic algorithms and greedy operators for graph bi-partitioning. *Evolutionary Computation*, Vol. 8, No. 1, pp. 61–91, 2000.

18. P. Merz and K. Katayama. Memetic algorithms for the unconstrained binary quadratic programming problem. *BioSystems*, Vol. 78, No. 1–3, pp. 99–118, 2004.

19. P. Moscato and C. Cotta. A gentle introduction to memetic algorithms. In F. Glover and G. Kochenberger, editors, *Handbook of Metaheuristics*, Vol. 57 of *International Series in Operations Research & Management Science*, pp. 105–144. Kluwer Academic Publishers, Norwell, MA, 2003.

20. P.R.J. Östergård. A fast algorithm for the maximum clique problem. *Discrete Applied Mathematics*, Vol. 120, No. 1–3, pp. 197–207, 2002.

21. W. Pullan and H.H. Hoos. Dynamic local search for the maximum clique problem. *Journal of Artificial Intelligence Research*, Vol. 25, pp. 159–185, 2006.

22. A. Singh and A. K. Gupta. A hybrid heuristic for the maximum clique problem. *Journal of Heuristics*, Vol. 12, No. 1-2, pp. 5–22, 2006.

23. E. Tomita and T. Seki. An efficient branch-and-bound algorithm for finding a maximum clique. In *Discrete Mathematics and Theoretical Computer Science, LNCS 2731*, pp. 278–289. Springer-Verlag, 2003.

Accelerating Local Search in a Memetic Algorithm for the Capacitated Vehicle Routing Problem

Marek Kubiak and Przemysław Wesołek

Institute of Computing Science, Poznan University of Technology
Piotrowo 2, 60-965 Poznan, Poland
Marek.Kubiak@cs.put.poznan.pl

Abstract. Memetic algorithms usually employ long running times, since local search is performed every time a new solution is generated. Acceleration of a memetic algorithm requires focusing on local search, the most time-consuming component. This paper describes the application of two acceleration techniques to local search in a memetic algorithm: caching of values of objective function for neighbours and forbidding moves which could increase distance between solutions. Computational experiments indicate that in the capacitated vehicle routing problem the usage of these techniques is not really profitable, because of cache management overhead and implementation issues.

1 Introduction

Population-based algorithms are usually more time-consuming than their single-solution-based counterparts. Evolutionary algorithms, as an example of the first type, employ large computation times, as compared to e.g. simulated annealing or tabu search. Memetic algorithms (MAs) [1], a kind of evolutionary ones, are even more prone to this problem; in a memetic algorithm a local search process is conducted for every solution in a population, which makes the process of computation even longer.

On the other hand, population-based algorithms usually offer the possibility of exploration of the search space to high extent and, thus, generate better solutions than procedures based solely on local search. Therefore, algorithms managing a population of solutions are a useful tool of optimization. Nevertheless, it would be profitable if the speed of memetic algorithms could be increased without deterioration in the quality of results.

The majority of computation time of a memetic algorithm is usually spent on local search, after each recombination and mutation [1]. Consequently, each attempt to speed up the whole algorithm should be focused on local search.

The main acceleration possibility in local search concerns the computation of quality of neighbours to a current solution. If the difference of the objective function between the current solution and its neighbour may be computed faster than the objective for the neighbour from scratch, then the whole process

C. Cotta and J. van Hemert (Eds.): EvoCOP 2007, LNCS 4446, pp. 96–107, 2007.
© Springer-Verlag Berlin Heidelberg 2007

speeds-up drastically. In [1] Merz claims that it is possible for almost every combinatorial optimization problem. Jaszkiewicz [2] mentions that local search for the TSP performed almost 300 times more function evaluations per second than a genetic procedure computing the objective from scratch. This is also the case in the capacitated vehicle routing problem (CVRP), which is considered here: neighbours of a solution (w.r.t. commonly used neighbourhood operators) may be evaluated quicker than random solutions.

However, Ishibuchi et al. [3] rightly note that there are problems for which such acceleration is not possible. They give an example of a flowshop problem with the completion time as the objective: a neighbour to a solution is not evaluated faster than a completely new solution. This results from the fact that certain objectives and/or constraints have global character and even a small change in contents of a solution require complete recomputation of the objective function and/or checking all constraints.

Another possibility of speeding-up local search requires caching of (storing in auxiliary memory) values of the difference in objective functions. This technique is not new and is also known as "don't look bits" [4]: if a neighbour of a solution has been evaluated as worse in a previous iteration of local search, then it is not evaluated at all in the current iteration. Such an approach requires that only the changing part of a neighbourhood of a current solution is evaluated.

The two mentioned techniques do not take the memetic search into account. However, there is a possibility to speed-up local search also based on information contained in the population of an MA. If the optimization problem considered exhibits the 'big valley' structure, then it means that good solutions of the problem are located near to each other, and to global optima, in the search space [5], [1], [6]. In such a case recombination operators of MAs should be respectful or distance-preserving [5], [7], [1]. Moreover, the local search process, which is always launched after a recombination, should also observe that the distance between an offspring and its parents is not inflated. This is the place where speed-up may be obtained: some moves of local search on an offspring should be forbidden and, therefore, some neighbours not checked for improvement at all, since they would lead to an increase in distance to parents. This technique was successfully applied in MAs by Merz [8] for the quadratic assignment problem and by Jaszkiewicz [9] for the TSP.

This paper is a study of the application of the two latter acceleration techniques to the capacitated vehicle routing problem. It firstly describes design of and experiments with cache in local search. Then, experiments with local search moves forbidden after distance-preserving recombination are presented. However, due to the nature of the analysed problem and implementation issues it appears that these techniques result in only small acceleration of the memetic algorithm.

2   The Capacitated Vehicle Routing Problem

The capacitated vehicle routing problem (CVRP) [10] is a very basic formulation of a problem which a transportation company might face in its everyday operations. The goal is to find the shortest-possible set of routes for the company's vehicles in order to satisfy demands of customers for certain goods. Each of identical vehicles starts and finishes its route at the company's depot, and must not carry more goods than its capacity specifies. All customers have to be serviced, each exactly once by one vehicle. Distances between the depot and customers are given.

The version of the CVRP considered here does not fix the number of vehicles (it is a decision variable); also the distance to be travelled by a vehicle is not constrained. Compared to the multiple-TSP, the CVRP formulates one more constraint, the capacity constraint: the sum of demands of customers serviced by one vehicle (i.e. in one route) must not exceed the vehicle's capacity.

Refer to [10] for more information about the CVRP.

3   Cache for Neighbourhood Operators

3.1   The Idea of Caching Evaluations of Neighbours

When applied to a solution, neighbourhood operators in local search for the CVRP usually modify only a small fragment of its contents. Large parts of this solution stay intact. Consequently, large number of moves which modified the original solution may also be performed for the modified, new one, and the modifications of the objective function stay the same. Therefore, there is no need to recompute this change of the objective; it may be stored in cache for later use.

Nevertheless, some moves from the original solution are changed by the actually performed move. These modified moves must not be stored; they have to be removed from the cache. The set of such moves strongly depends on the performed move.

These remarks lead to the following algorithm of local search with cache:

$localSearch(s)$
    do:
        for each $s' \in N(s)$ do:
            if $\Delta f(s', s)$ is stored in the cache:
                $\Delta f = \Delta f(s', s)$ is taken from the cache
            else:
                compute $\Delta f = \Delta f(s', s) = f(s') - f(s)$
                store $\Delta f(s', s)$ in the cache for later use
            if $\Delta f < 0$ then $s_i = s'$ is an improved neighbour of s
        if s_i has been found (an improved neighbour of s):
            $s = s_i$ (move to the neighbour)
            update the cache:
                for each $s_a \in N(s)$ affected by the move, delete $\Delta f(s_a, s)$ from
                    the cache
        else: break the main loop (a local optimum was found)
      while (true)
      return s (a local optimum)

From this description one may notice the possible source of gain in speed: instead of computing $\Delta f(s', s) = f(s') - f(s)$ (the fitness difference) for each neighbour s' of s, this value is stored in the cache for later use. However, the operation of cache update, which has to be called after a move is found in order to ensure the cache stays valid, is a possible source of computation cost. The goal of caching is to make the gain higher than the cost.

Local search is usually utilised in one of two possible ways: first improvement (greedy) or best improvement (steepest). It may be predicted [11] that the gain from caching will be greater for the steepest algorithm. It has to check the whole neighbourhood in every iteration, so the auxiliary memory will be fully up-to-date. In case of the greedy algorithm cache is initially empty and stays in this state for many iterations, until it becomes hard to find an improving neighbour. Only then it is filled with up-to-date values. However, the overhead connected with cache updates is present in every iteration.

3.2   Cache Requirements

In the CVRP not only the objective function matters. There is also the capacity constraint, which involves whole routes, not only single customers. Thus, if the capacity constraint for a neighbour is violated then this neighbour is infeasible; such moves are forbidden in local search. Therefore, not only the change in the objective function has to be stored in the cache, but also the status of feasibility of a neighbour.

Three neighbourhood operators are considered here (size of a neighbourhood is given in brackets):

- *merge*: merge of any 2 routes ($O(T^2)$; T is the number of routes in a solution)
- *2opt*: exchange of any 2 edges ($O((n + T)^2)$; n is the number of customers)
- *swap*: exchange of any 2 customers ($O(n^2)$)

Because these operators have different semantics, cache must be designed and implemented independently for each of them (in separate data structures).

The local search considered here assumes that the neighbourhoods of these operators may be joined to form one large neighbourhood. It also means that the order of execution of operators cannot be determined in advance (it may be e.g.: *merge, merge, 2opt, swap, 2opt,...*; it may be any other order). Such a possibility makes local search potentially more powerful (there are less local optima in the search space) but also more time-consuming. In case of cache this possibility creates a requirement that when one type of move is performed, then cache of all operations has to be updated.

The neighbourhoods of the operators have different sizes; the neighbourhood of *2opt* and *swap* is considerably larger than the one of *merge*. Moreover, the *merge* operation is very specific: the number of applications of this operator is always very limited by the minimum possible number of routes. Finally, initial experiments with MAs indicated that the number of applications of this operator amounts to 5–10% of the total number of applications of all operators; the majority of search effort is spent on *2opt* and *swap*. Therefore, the cache was implemented for these two operators only. The size of memory for the cache structures is the same as the size of the related neighbourhoods.

4   Speeding-Up *2opt* Feasibility Checks

In the CVRP, *2opt* may be used in two main configurations [12]:

- exchanging 2 edges inside one route (in-route *2opt*),
- exchanging 2 edges between two different routes (between-routes *2opt*).

The main computation cost of finding an improving *2opt* move is related to feasibility checks of between-routes *2opt*; it involves two routes, which may become infeasible after the move is performed, due to the capacity constraint present in the CVRP.

For the exemplary solution shown in Figure 1 (top, left) there are two ways in which a *2opt* may be executed if removing edges $(2, 3)$ and $(8, 9)$ (the marked ones):

- by connecting $(2, 8)$ and $(3, 9)$ (Figure 1, top, centre);
- by connecting $(2, 9)$ and $(3, 8)$ (Figure 1, top, right).

Both of these between-routes *2opt* configurations are prone to infeasibility; e.g. while connecting $(2, 8)$ and $(3, 9)$ if the sum of demands of customers $(1, 2, 8, 7)$ or $(6, 5, 4, 3, 9, 10, 11, 12)$ exceeds the capacity, then this move is infeasible.

All such moves require, therefore, that parts of routes (e.g. the mentioned $(1, 2)$ and $(8, 7)$) have known demands, so they could be added for the feasibility check. This is the cause of additional high computation cost in local search (pessimistically: $O(n)$), if these parts of demands are computed from scratch every time a between-routes *2opt* is checked.

In [12] a technique was described which reduces this cost to a constant. It is based on the observation that demands of parts of routes may be stored and simply updated when iterating over neighbours of a current solution in a right order. This order is called a lexicographic one.

An example of such order is given in Figure 1. The top of Figure 1 shows a *2opt* removing edges $(2, 3)$ and $(8, 9)$ (as described above); the demands of parts $(1, 2)$, $(3, 4, 5, 6)$, $(7, 8)$, $(9, 10, 11, 12)$ are required. The bottom of Figure 1 shows the immediately next *2opt* moves (in the lexicographic order), the ones removing edges $(2, 3)$ and $(9, 10)$. The required demands of parts $(1, 2)$, $(3, 4, 5, 6)$ have just been computed in the previous iteration and may be used; the demands of parts $(7, 8, 9)$ and $(10, 11, 12)$ may be computed from the previous values at the cost of two additions.

Due to the high predicted gain in computation time, this technique was used in local search in each configuration with cache.

5   Forbidden Moves of Local Search After Recombination

Based on the results of 'big valley' examination in the CVRP [7] it is known that preservation of edges is important for quality of solutions, as it is the case of the

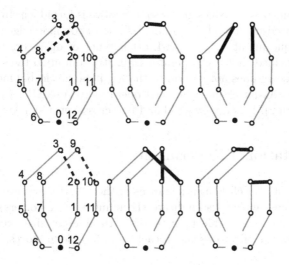

Fig. 1. Edge exchanges in lexicographic order: *2opt* for $(2,3)$, $(8,9)$ (*top*) and *2opt* for $(2,3)$ and $(9,10)$ (*bottom*)

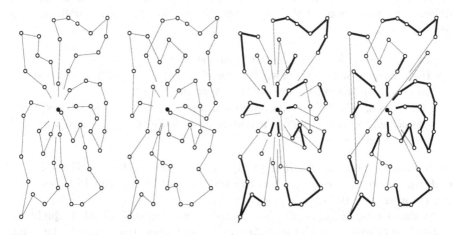

Fig. 2. Example of application of crossover operators: parent1, parent2, CECPX2 offspring (common edges emphasized), SPX offspring (edges from parent1 emphasized)

TSP [9]. Therefore, Kubiak [7] proposed a set of distance-preserving crossover operators for the problem. One of them, CECPX2, creates an offspring in such a way that it always contains all edges common to both parents (common edges), possibly including some additional ones. An example of application of CECPX2 is shown in Figure 2.

Having the idea of 'big valley' and distance preservation in mind, it makes sense after CECPX2 to forbid in local search all moves which would change any edge from the set of common edges. Therefore, an offspring of CECPX2 has the common edges marked as 'forbidden'. All neighbourhood operators check if a move deleted one of such edges and if so, the move is forbidden. In consequence, a significant speed-up should be obtained if sets of common edges are large.

This technique might be used with other types of operators (not distance-preserving) provided that such operator explicitly computed the set of common edges. If a crossover does not determine the set, it cannot forbid moves changing common edges. As an example of such operator we use SPX (its offspring is shown in Figure 2). It is a very good and fast crossover designed by Prins [13].

6   Computational Experiments

In all experiments 7 well-known instances of the CVRP were used, taken from [14]. Their names (which also indicate the number of customers) are given in tables with results, e.g. Table 1. Instances with different sizes were selected in order to observe the effect of scale in cache and forbidden moves.

6.1   Experiments with Local Search

In order to asses the efficiency of cache in local search, an experiment with 10 different configurations of this algorithm was conducted. These configurations resulted from:

- two versions of local search: greedy and steepest;
- three versions w.r.t. cache: without cache (denoted nc); with all cache structures (c); the same as c, but without *2opt* cache (c*);
- two types of neighbourhoods: one joined neighbourhood of *merge, 2opt, swap* (described as n-3); a *merge* neighbourhood followed by a joined *2opt* and *swap* (described as n-1-2).

Each configuration was run in a multiple start local search (MSLS) algorithm, each time starting from a new random solution. MSLS was run 10 times; each run stopped after 100 LS processes.

Average times of computation in this experiment are given in Table 1. Quality of results is not given, since all the configurations had them approximately the same.

The greedy version of local search is several times faster than the steepest one. The version n-1-2 of greedy search is slightly faster than n-3. These results were expected: greedy is usually faster; n-1-2 searches smaller neighbourhoods.

What is more important, the usage of cache in n-3 drastically deteriorates the runing times. This might be explained by the *merge* operations included in the neighbourhood. This operation has no cache on its own, but each time it is performed it results in updates to cache of other operators, making the cache almost empty and the cache management cost unacceptably high.

The usage of cache in n-1-2 gives no improvement, as well.

For the steepest version of local search, the comparison of n-3 and n-1-2 (no cache) yields the same conclusions: the latter is faster. The cache also deteriorates the situation here. Only the c* version results in slight improvements.

Table 1. Average times of computation (in seconds) for local search algorithms, greedy (*left*) and steepest (*right*)

	n-3		n-1-2				n-3		n-1-2		
instance	nc	c	nc	c	c^*	instance	nc	c	nc	c	c^*
c50	0.7	2.0	0.7	1.2	0.5	c50	3.5	5.0	2.0	2.7	2.0
tai75d	2.5	6.9	2.2	4.1	2.5	tai75d	14.2	18.4	8.8	10.1	8.1
tai100d	5.5	16.8	4.8	8.4	5.5	tai100d	34.0	44.3	20.4	22.6	18.5
c120	11.5	34.6	8.7	14.3	9.2	c120	61.3	80.1	36.8	39.4	31.0
tai150b	20.8	83.1	17.9	27.2	17.7	tai150b	122.7	178.7	77.5	81.2	68.8
c199	30.4	289.4	29.1	43.3	29.5	c199	279.4	519.9	168.8	173.5	147.1
tai385	299.9	5649.1	301.3	394.3	292.3	tai385	2543.5	7897.6	1597.5	1594.8	1385.7

To summarize, the results of this experiment show that cache structures do not really improve LS running times. Instead, they slow LS down in many configurations. The cause of this effect lies most probably in the capacity constraint. Because of this constraint the operation of cache updates is time-consuming: if a move to a neighbour changes more than one route (which happens often with *2opt* and *swap*), then a large part of cache has to be invalidated – all moves concerning every part of the modified routes. This is not the case e.g. in TSP or other unconstrained problems (see [1]).

The possible source of this disappointing result might also lie in details of cache implementation.

6.2 Local Search Execution Profiles

We decided to make detailed profiles of local search executions in order to gain insights into the cost of search and cache operations. In this case, analytical computation of cost is difficult: it is hard to compute the actual or expected number of local search iterations, or to estimate the cache usage. That is why we analysed the issue empirically [11].

We tested LS with the following settings:

- LS version: greedy or steepest,
- cache usage: without cache (nc); with all cache structures (c); with the most promising cache settings, leaving *2opt* cache out as too costly (c^*),
- neighbourhood: n-1-2.

Only two instances were tested, tai100d and c120. One run of MSLS was conducted for each setting and instance, consisting of 5 independent LS processes. The runs were limited and short because code profiling usually considerably increases the run time due to injection of timing routines into the original code.

The profiling results of greedy LS for instance c120 are presented in Table 2. They contain the times of operations (search and cache) for each profiled LS setting. Also percentages of the total run time of the base version (n-1-2-nc) are

Table 2. Times of execution of search and cache operations in greedy LS; c120

operation	n-1-2-nc		n-1-2-c		n-1-2-c*	
	time [s]	(percent)	time [s]	(percent)	time [s]	(percent)
2opt: eval. of neighbours	146.3	(50.4)	47.6	(16.4)	57.1	(19.7)
2opt: cache read/write	0.0	(0.0)	30.1	(10.4)	2.6	(0.9)
2opt: cache update	0.0	(0.0)	19.8	(6.8)	0.0	(0.0)
2opt: total search cost	146.3	(50.4)	97.5	(33.6)	59.7	(20.6)
swap: eval. of neighbours	33.2	(11.4)	13.3	(4.6)	11.5	(4.0)
swap: cache read/write	0.0	(0.0)	8.1	(2.8)	7.2	(2.5)
swap: cache update	0.0	(0.0)	5.1	(1.8)	5.1	(1.8)
swap: total search cost	33.2	(11.4)	26.5	(9.2)	23.8	(8.2)
operators: total	179.5	(61.8)	124.0	(42.7)	83.5	(28.8)
greedy LS: total	290.2	(100.0)	209.9	(72.3)	173.1	(59.6)

shown. The *merge* operator is not reported due to insignificant cost of its operations (1–2% of the total run time in all runs).

For the greedy version without cache, very high cost of search by the *2opt* operator is clearly visible (50.4% of the total run time). The cost of *swap* is lower, although it is also considerable (11.4%). Consequently, there is space for improvement in this base version.

The introduction of cache decreases the *2opt* evaluation time, to 16.4%. However, it introduces new cost components: cache reads and writes (10.4%), and cache updates after a performed move (6.8%). In total, the *2opt* search time drops from 50.5% to 33.6%; it seems that the decrease is not as high as could be: the *2opt* cache cost is considerable.

The same conclusion applies to the *swap* operator: the evaluation time drops from 11.4% to 4.6%, but cache management (read/write and update) takes another 4.6%, making the cache only slightly profitable.

The analysis of cache usage in these profiled runs demonstrated that only 28.1% of *2opt* cache is used, while for *swap* it is 58.8%. As predicted, the cached values are rarely used in the greedy version, because improving steps are usually found very quickly (the neighbourhood is not completely searched through, sparsely filling cache with valid values). Moreover, these numbers indicate that *2opt* updates invalidate large parts of cache, while for the *swap* operator most of the cache stays valid after an improving move is performed.

The last setting, n-1-2-c*, did not use *2opt* cache; it seemed that the cache management cost for this operator was too high. The results show that this approach gives the highest gain for the greedy LS: the evaluation cost for *2opt* amounts to 19.7%, but the management cost is almost none (the figure 0.9% reflects the time of calls to empty cache which is not updated at all). In conjunction with some gain from *swap*, the overall speed-up of LS equals 40.4%.

In case of instance tai100d (the detailed results are not reported) the cache management cost was generally higher, making cache usage too expensive. It indicates that cache may be beneficial for larger instances only, if ever.

The steepest version differed mainly in the cache usage: 62.0% of *2opt* neighbours were evaluated based on the cache contents; as much as 77.2% in the case of *swap*. Therefore, gains from cache were slightly higher.

To summarise this experiment, the execution profiles indicate high cache management cost which is generally compensated by gain in evaluation of neighbours, but results in no further significant improvement.

6.3   Experiments with a Memetic Algorithm

The evaluation of forbidden changes (FC) required an experiment with memetic algorithm. This algorithm was run in 12 different configurations renderred by:

- two versions of embedded local search: greedy and steepest;
- two versions w.r.t. cache usage in local search: nc (no cache) and c* (cache, but without *2opt*);
- three variants of crossover: SPX, CECPX2 and CECPX2 with forbidden changes (CECPX2-FC).

Each configuration was run 30 times. The stop criterion was the total number of generations, equal to the average number of generations required for the MA to converge. The algorithms running times are gathered in Table 3.

Table 3. Average times of computation (in seconds) for memetic algorithms, greedy (*above*) and steepest (*below*)

instance	SPX		CECPX2		CECPX2-FC	
	nc	c*	nc	c*	nc	c*
c50	1.9	2.0	1.0	1.0	1.0	1.0
tai75d	13.9	13.9	12.6	12.4	12.7	12.5
tai100d	31.2	30.6	29.4	28.7	29.7	29.1
c120	67.2	57.0	67.5	56.3	68.9	56.6
tai150b	234.2	217.9	218.6	206.9	221.1	206.6
c199	352.7	317.9	373.0	343.9	376.6	341.7
tai385	4170.3	3695.7	5033.2	4377.3	4990.6	4387.5

instance	SPX		CECPX2		CECPX2-FC	
	nc	c*	nc	c*	nc	c*
c50	2.0	2.0	2.0	2.0	2.0	2.0
tai75d	15.7	14.4	12.3	11.3	12.3	11.2
tai100d	36.1	32.2	30.0	27.8	29.7	26.2
c120	61.4	49.8	56.7	47.9	56.9	47.7
tai150b	201.5	181.0	173.9	163.7	178.4	162.3
c199	318.5	277.3	301.0	274.2	297.3	261.9
tai385	3264.5	2767.5	2897.1	2623.0	2879.8	2584.6

For greedy versions of local search embedded in the memetic algorithm the configuration with operator SPX is the fastest, especially with cache (c^*), which results in some small gain in computation time (approx. 10%). For the CECPX2 operator also the versions with cache are a bit faster (also approx. 10%). The forbidden changes gain nothing, though.

The MAs with steepest local search are significantly faster than their greedy counterparts. Similarly to the latter, cache usage intoduces a small speed-up. For steepest configurations, CECPX2 is generally faster than SPX. The application of forbidden changes gives in effect a tiny gain in computation time.

In summary, however, it has to be said that both of the applied techniques, cache and forbidden changes, did not provide the expected acceleration in the memetic algorithm.

7   Conclusions

The paper presented experiments with some acceleration techniques for local search and memetic algorithm applied to the CVRP.

The obtained results indicate that the application of cache in the LS gives no real gain; compared to results reported for other problems (e.g. the TSP) there is no profitability in using cache. It seems that the main problem in the cache for the CVRP is the cache management cost, which results from the need to update the cache contents each time an improving move is performed (the capacity constraint forces large parts of cache to be invalidated). The application of forbidden moves in the MA also leads to no gain.

Comparing roughly the implementation cost it appears that cache design and implementation is very expensive (especially the tests of correctness of cache updates), which makes it an inefficient technique. The cost of implementation of forbidden changes was, in contrast, surprisingly low. Perhaps this technique requires some more attention from the authors and it may in future lead to some improvement.

However, the comparison of results from this and the initial version of this paper revealed that the running times of both LS and MA decreased 5–10 times in the final version. It was the result of changes in parts of local search code which affected all the configurations analysed in this paper. These were low-level, implementation changes (e.g. method inlining, avoiding calls to copying constructors by passing references, etc.) introduced after first code profiling. These changes caused the gains from cache and forbidden changes to become virtually invisible, although they were noticeable in first experiments. Perhaps the implementation of cache operations was not as efficient as it could be. Further implementation work should resolve this issue.

Finally, the authors are satisfied with acceleration of local search and memetic algorithm achieved during preparation of this paper.

Acknowledgements. This work was supported by Polish Ministry of Science and Higher Education through grant no 8T11F00426.

References

1. Merz, P.: Advanced fitness landscape analysis and the performance of memetic algorithms. Evolutionary Computation **12**(3) (2004) 303–325
2. Jaszkiewicz, A.: Genetic local search for multiple-objective combinatorial optimization. European Journal of Operational Research **137**(1) (2002) 50–71
3. Ishibuchi, H., Yoshida, T., Murata, T.: Balance between genetic search and local search in memetic algorithms for multiobjective permutation flowshop scheduling. IEEE Transactions on Evolutionary Computation **7**(2) (2003) 204–223
4. Bentley, J.L.: Experiments on traveling salesman heuristics. In: Proceedings of the first annual ACM-SIAM symposium on discrete algorithms. (1990) 91–99
5. Jaszkiewicz, A., Kominek, P.: Genetic local search with distance preserving recombination operator for a vehicle routing problem. European Journal of Operational Research **151**(2) (2003) 352–364
6. Reeves, C.R.: Landscapes, operators and heuristic search. Annals of Operations Research **86**(1) (1999) 473–490
7. Kubiak, M.: Systematic construction of recombination operators for the vehicle routing problem. Foundations of Computing and Decision Sciences **29**(3) (2004) 205–226
8. Merz, P., Freisleben, B.: A genetic local search approach to the quadratic assignment problem. In Bäck, T., ed.: Proceedings of the Seventh International Conference on Genetic Algorithms. (1997)
9. Jaszkiewicz, A.: Improving performance of genetic local search by changing local search space topology. Foundations of Computing and Decision Sciences **24**(2) (1999) 77–84
10. Toth, P., Vigo, D.: The Vehicle Routing Problem. SIAM, Philadelphia (2002)
11. Hoos, H.H., Stutzle, T.: Stochastic Local Search: Foundations and Applications. Morgan Kauffman (2004)
12. Kindervater, G.A.P., Savelsbergh, M.W.P.: Vehicle routing: handling edge exchanges. In Aarts, E., Lenstra, J.K., eds.: Local Search in Combinatorial Optimization. John Wiley & Sons (1997) 337–360
13. Prins, C.: A simple and effective evolutionary algorithm for the vehicle routing problem. In de Sousa, J.P., ed.: Proceedings of MIC 2001, the 4th Metaheuristics International Conference. (2001) 143–147
14. Rochat, Y., Taillard, É.D.: Probabilistic diversification and intensification in local search for vehicle routing. Journal of Heuristics **1**(1) (1995) 147–167

Evolutionary Algorithms for
Real-World Instances of the Automatic
Frequency Planning Problem in GSM Networks*

Francisco Luna[1], Enrique Alba[1], Antonio J. Nebro[1], and Salvador Pedraza[2]

[1] Department of Computer Science, University of Málaga (Spain)
{flv,eat,antonio}@lcc.uma.es
[2] Optimi Corp., Edif. Inst. Universitarios, Málaga (Spain)
Salvador.Pedraza@optimi.com

Abstract. Frequency assignment is a well-known problem in Operations Research for which different mathematical models exist depending on the application specific conditions. However, most of these models are far from considering actual technologies currently deployed in GSM networks (e.g. frequency hopping). These technologies allow the network capacity to be actually increased to some extent by avoiding the interferences provoked by channel reuse due to the limited available radio spectrum, thus improving the Quality of Service (QoS) for subscribers and an income for the operators as well. Therefore, the automatic generation of frequency plans in real GSM networks is of great importance for present GSM operators. This is known as the Automatic Frequency Planning (AFP) problem. In this paper, we focus on solving this problem for a realistic-sized, real-world GSM network by using Evolutionary Algorithms (EAs). To be precise, we have developed a $(1, \lambda)$ EA for which very specialized operators have been proposed and analyzed. Results show that this algorithmic approach is able to compute accurate frequency plans for real-world instances.

1 Introduction

The *Global System for Mobile* communication (GSM) [1] is an open, digital cellular technology used for transmitting mobile voice and data services. GSM is also referred to as 2G, because it represents the second generation of this technology, and it is certainly the most successful mobile communication system. Indeed, by mid 2006 GSM services are in use by more than 1.8 billion subscribers[1] across 210 countries, representing approximately 77% of the world's cellular market. It is widely accepted that the *Universal Mobile Telecommunication System* (UMTS) [2], the third generation mobile telecommunication system, will coexist with the enhanced releases of the GSM standard (GPRS [3] and EDGE [4])

---

* This work has been partially funded by the Ministry of Science and Technology and FEDER under contract TIN2005-08818-C04-01 (the OPLINK project).

[1] http://www.wirelessintelligence.com/

C. Cotta and J. van Hemert (Eds.): EvoCOP 2007, LNCS 4446, pp. 108–120, 2007.

at least in the first phases. Therefore, GSM is expected to play an important role as a dominating technology for many years.

The success of this multi-service cellular radio system lies in efficiently using the scarcely available radio spectrum. GSM uses *Frequency Division Multiplexing* and *Time Division Multiplexing* schemes to maintain several communication links "in parallel". The available frequency band is slotted into channels (or frequencies) which have to be allocated to the elementary transceivers (TRXs) installed in the base stations of the network. This problem is known as the Automatic Frequency Planning (AFP) problem, the Frequency Assignment Problem (FAP), or the Channel Assignment Problem (CAP). Several different problem types are subsumed under these general terms and many mathematical models have been proposed since the late sixties [5,6,7]. This work is focussed on concepts and models which are relevant for current GSM frequency planning [8] and not on simplified models of the abstract problem. In GSM, a network operator has usually a small number of frequencies (few dozens) available to satisfy the demand of several thousands TRXs. A reuse of these frequencies is therefore unavoidable. However, frequency reuse is limited by interferences which could lead the quality of service (QoS) for subscribers to be reduced to unsatisfactory levels. Consequently, the automatic generation of frequency plans in real GSM networks is a very important task for present GSM operators not only in the initial deployment of the system, but also in subsequent expansions/modifications of the network, solving unpredicted interference reports, and/or handling anticipate scenarios (e.g. an expected increase in the traffic demand in some areas). Additionally, several interference reduction techniques (e.g. frequency hopping, discontinuous transmission, or dynamic power control) [8] have been proposed to enhance the capacity of a given network while using the same frequency spectrum. These techniques are currently in use in present GSM networks and they must be carefully considered in AFP problems because they allow both the QoS for subscribers and the income of the operators to be increased.

The AFP problem is a generalization of the graph coloring problem, and thus it is NP-hard [9]. As a consequence, using exact algorithms to solve real-sized instances of AFP problems is not practical, and therefore other approaches are required. Many different methods have been proposed in the literature [5] and, among them, metaheuristic algorithms have proved to be particularly effective. Metaheuristics [10,11] are stochastic algorithms that sacrifice the guarantee of finding optimal solutions for the sake of (hopefully) getting accurate (also optimal) ones in a reasonable time. This fact is even more important in commercial tools, in which the GSM operator cannot wait for long times to get a frequency plan (e.g. several weeks). Our approach here is to use Evolutionary Algorithms (EAs) [12]. However, it has been reported in the literature that classical EA crossover operators do not work properly for this problem [13,14]. Our proposal is therefore a fast and accurate $(1, \lambda)$ EA (see [15] for details on this notation) in which there is no need for recombining individuals. A $(1, \lambda)$ EA is an approach that either shows population-based evolutionary capabilities and a low cost per iteration (similar to Simulated Annealing and other trajectory based algorithms). These two are the reasons for choosing this algorithm instead of a regular (μ, λ) EA (like regular GAs) of (1,1) EA (like greedy approaches). The main contribution of this work is not only using a real-world GSM network instance with real data and realistic size (more than 2600 TRXs to be assigned just 18 frequencies) but also that the tentative frequency plans manipulated by the EA are evaluated with a commercial tool which uses accurate models for all the system components (signal propagation, TRX, locations, etc.) and actually deployed GSM interference reduction technologies such as those mentioned above. Both the data as well as the simulator are provided by Optimi Corp.™ . The point here is that standard benchmarks like the Philadelphia instances, CELAR, and COST 259 [6] do not consider such technologies and therefore most of the proposed optimization algorithms are rarely faced with a real GSM frequency planning problem. We have implemented specialized operators for the $(1, \lambda)$ EA in which precise network information has been used. Finally, different configurations of the algorithm ($\lambda = 10$ and $\lambda = 20$) showing different balances between intensification/diversification have been tested. The results point out that our approach is able to compute accurate frequency plans that can be directly deployed in the real GSM network used.

The paper is structured as follows. In the next section, we provide the reader with details on the frequency planning in GSM networks. Section 3 describes the algorithm proposed along with the different operators used. The results of the experiments are analyzed in Section 4. Finally, conclusions and future lines of research are discussed in the last section.

2   Frequency Planning in GSM Networks

This section is devoted to presenting details on the frequency planning task for a GSM network. We first provide the reader with a brief description of the GSM architecture. Next, we give the relevant concepts to the frequency planning problem that will be used along this paper.

2.1   The GSM System

An outline of the GSM network architecture is shown in Fig. 1. As it can be seen, GSM networks are built out of many different components. The most relevant ones to frequency planning are the Base Transceiver Station (BTS) and the transceivers (TRXs). Essentially, a BTS is a set of TRXs. In GSM, one TRX is shared by up to eight users in TDMA (*Time Division Multiple Access*) mode. The main role of a TRX is to provide conversion between the digital traffic data on the network side and radio communication between the mobile terminal and the GSM network. The site at which a BTS is installed is usually organized in sectors: one to three sectors are typical. Each sector defines a cell.

The solid lines connecting components in Fig. 1 carry both traffic information (voice or data) as well as the "in-band" signaling information. The dashed lines are signaling lines. The information exchanged over these lines is necessary

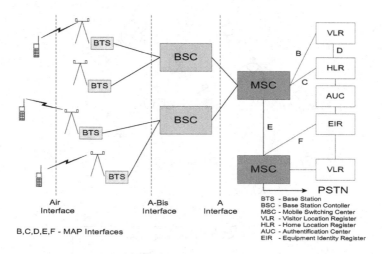

Fig. 1. Outline of the GSM network architecture

for supporting user mobility, network features, operation and maintenance, authentication, encryption, and many other functions necessary for the network's proper operation. Fig. 1 shows the different network components and interfaces within a GSM network.

2.2   The Automatic Frequency Planning Problem

The frequency planning is the last step in the layout of a GSM network. Prior to tackling this problem, the network designer has to address some other issues: where to install the BTSs or how to set configuration parameters of the antennas (tilt, azimuth, etc.), among others [16]. Once the sites for the BTSs are selected and the sector layout is decided, the number of TRXs to be installed per sector has to be fixed. This number depends on the traffic demand which the corresponding sector has to support. The result of this process is a quantity of TRXs per cell. A channel has to be allocated to every TRX and this is the main goal of the AFP [8]. Essentially, three kinds of allocation exist: Fixed Channel Allocation (FCA), Dynamic Channel Allocation (DCA), and Hybrid Channel Allocation. In FCA, the channels are permanently allocated to each TRX, while in DCA the channels are allocated dynamically upon request. Hybrid Channel Allocation schemes (HCA) combine FCA and DCA. Neither DCA nor HCA are supported in GSM, so we only consider FCA.

We now explain the most important parameters to be taken into account in GSM frequency planning. Let us consider the example network shown in Fig. 2, in which each site has three installed sectors (e.g. site A operates $A1$, $A2$, and $A3$). The first issue is the implicit topology which results from the previous steps in the network design. In this topology, each sector has an associated list of neighbors containing the possible handover candidates for the mobile residing in a specific cell. These neighbors are further distinguished into first order (those which can

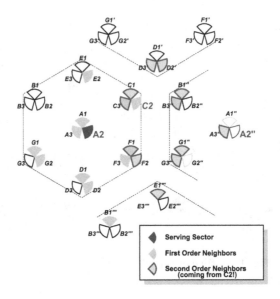

Fig. 2. An example of GSM network

potentially provoke strong interference to the serving sector) and second order neighbors. In Fig. 2, $A2$ is the serving sector and the first order neighbors defined are $A1$, $A3$, $C2$, $D1$, $D2$, $E2$, $F3$, $G1$, $G2$, and $B1'''$, whereas, if we consider $C2$, second order neighbors of $A2$ are $F1$, $F2$, $C1$, $C3$, $D2'$, $D3'$, $A3''$, $B1''$, $B3''$, $G1''$, $G3''$, and $E1'''$.

As stated before, each sector in a site defines a cell; the number of TRXs installed in each cell depends on the traffic demand. A valid channel from the available spectrum has to be allocated to each TRX. Due to technical and regulatory restrictions, some channels in the spectrum may not be available in every cell. They are called locally blocked and they can be specified for each cell.

Each cell operates one Broadcast Control CHannel (BCCH), which broadcasts cell organization information. The TRX allocating the BCCH can also carry user data. When this channel does not meet the traffic demand, some additional TRXs have to be installed to which new dedicated channels are assigned for traffic data. These are called Traffic CHannels (TCHs).

In GSM, significant interference may occur if the same or adjacent channels are used in neighboring cells. Correspondingly, they are named co-channel and adj-channel interference. Many different constraints are defined to avoid strong interference in the GSM network. These constraints are based on how close the channels assigned to a pair of TRXs may be. These are called separation constraints, and they seek to ensure the proper transmission and reception at each TRX and/or that the call handover between cells is supported. Several sources of constraint separation exists: co-site separation, when two or more TRXs are installed in the same site, or co-cell separation, when two TRXs serve the same cell (i.e., they are installed in the same sector).

This is intentionally an informal description of the AFP problem in GSM networks. It is out the scope of this work to propose a precise model of the problem, since we use a proprietary software which is aware of all these concepts, as well as the consideration of all the existing interference reduction techniques developed for efficiently using the scarce frequency spectrum available in GSM.

3   EAs for Solving the AFP Problem

EAs have been widely used for solving many existing flavors of the frequency assignment problem [5,6,8,17]. However, it has been shown that well known crossover operators such as single point crossover do not perform well on this problem [13]. Indeed, it does not make sense for a frequency plan to randomly exchange two different, possibly non-related assignments. Our approach here is to use an $(1, \lambda)$ Evolutionary Algorithm, in which the recombination operator is not required. Next, we first describe the generic (μ, λ) EA. The solution encoding used, the fitness function, the method used for generating the initial solutions, and several proposals for perturbing individuals are discussed afterwards.

3.1 (μ, λ) Evolutionary Algorithm

This optimization technique firstly generates μ initial solutions. Next, the algorithm perturbs and evaluates these μ individuals at each iteration, from which λ new ones are obtained. Then, the best μ solutions taken from the newly generated λ individuals are moved to the next iteration (note that μ is not bigger than λ). An outline of the algorithm is shown in Fig. 3. Other works using a similar algorithmic approach for the AFP problem can be found in [13,18].

As stated before, the configurations used in this work employ a value of $\mu = 1$. The seeding procedure for generating the initial solution and the perturbation operator are the core components defining the exploration capabilities

```
1:  P = new Population(μ);
2:  PAux = new Population(λ);
3:  init(P);
4:  evaluate(P);
5:  for iteration = 0 to NUMBER_OF_ITERATIONS do
6:    for i = 1 to λ do
7:      individual = select(P);
8:      perturbation = perturb(individual);
9:      evaluate(perturbation);
10:     PAux = addTo(PAux,perturbation);
11:   end for
12:   P = bestIndividuals(PAux,μ);
13: end for
```

Fig. 3. Pseudocode of the (μ, λ) EA

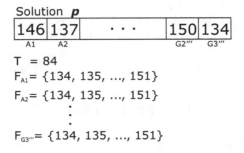

Fig. 4. Solution encoding example

of the $(1, \lambda)$ EA. The definition of these two procedures is detailed below in Sections 3.4 and 3.5.

3.2 Solution Encoding

The solution encoding determines both the search space and the subsequent set of search operators that can be applied during the exploration of this search space. Let T be the number of TRXs needed to meet the traffic demand of a given GSM network. Each TRX has to be assigned a channel. Let $F_i \subset \mathbb{N}$ be the set of available channels for the transceiver i, $i = 1, 2, 3, \ldots, T$. A solution p (a frequency plan) is encoded as a T-length integer array $p = [f_1, f_2, f_3, \ldots, f_T]$, $p \in F_1 \times F_2 \times \cdots \times F_T$, where $f_i \in F_i$ is the channel assigned to TRX i. The fitness function (see next section) is aware of adding problem specific information to each transceiver, i.e., whether it allocates either a BCCH channel or a TCH channel, whether it is a frequency hopping TRX, etc.

As an example, Fig. 4 displays the representation of a frequency plan p for the GSM network shown in Fig. 2. We have assumed that the traffic demand in the example network is fulfilled by one single TRX per sector (TRX $A1$, TRX $A2$, etc.).

3.3 Fitness Function

As it was stated before, we have used a proprietary application provided by Optimi Corp.™, that allows us to estimate the performance of the tentative frequency plans generated by the evolutionary optimizer. Factors like Frame Erasure Rate, Block Error Rate, RxQual, and BER are evaluated. This commercial tool combines all aspects of network configuration (BCCHs, TCHs, etc.) including interference reduction techniques (frequency hopping, discontinuous transmission, etc.) in a unique cost function, C, which measures the impact of proposed frequency plans on capacity, coverage, QoS objectives, and network expenditures. This function can be roughly defined as:

$$C = \sum_{v} \left(CostIM\left(v\right) \cdot E\left(v\right) + CostNeighbor\left(v\right) \right), \tag{1}$$

that is, for each sector v which is a potential victim of interference, the associated cost is composed of two terms, a signaling cost computed with the interference matrix $(CostIM\,(v))$ that is scaled by the traffic allocated to v, $E\,(v)$, and a cost coming from the current frequency assignment in the neighbors of v. Of course, the lower the total cost the better the frequency plan, i.e., this is a minimization problem.

3.4 Solution Initialization

Individuals are initialized site by site using a constructive method. For each site in the GSM network, a hopefully optimal frequency assignment is heuristically computed independently and without taking into account possible interferences from any other site. A simple greedy heuristic [5] is the method used (see Fig. 5 for its pseudocode). Given a site s, all its TRXs installed are randomly ranked (line 3). Then, random frequencies are assigned to the TRXs so that neither co-channel nor adjacent-channel interferences are provoked (lines 5 and 6). We want to note that the available radio spectrum is large enough to generate optimal frequency assignments within a site most times, i.e., we are applying the greedy heuristic independently only to the TRXs within a site. This way we avoid the most important source of strong interference through the network: those involving TRXs installed in the same site or sector.

```
1:   trxs = frequencies = ∅;
2:   trxs = TRXsFromSite(s);
3:   random_shuffle(trxs);
4:   for t in trxs do
5:       f = chooseInterferenceFreeFrequency(t,frequencies);
6:       assign(t,f);
7:       frequencies = insert(frequencies,t);
8:   end while
```

Fig. 5. Pseudocode of a greedy heuristic

Finally, the individual undergoes a heuristic which is based on the *Dsatur with Costs* heuristic [8]. Here, all the TRXs of the network are ranked so the hardest ones to deal with are assigned first. The measure used for "hardest to deal with" consists of accurate information given by the simulator about the importance of the TRXs (interference provoked, capacity overload, etc.), rather than using the generalization of saturation and space degrees of *Dsatur with Costs*. Next, each TRX is assigned the frequency presently incurring the least additional interference. The main goal of this phase is to reach "good" solutions in short times. This is usually a requirement within commercial applications, the context of this work. Using this greedy phase is also the main reason to use the $(1, \lambda)$ strategy. Indeed, this greedy step leads the search towards a local minimum and it makes the non-elitist $(1, \lambda)$ strategy suitable for the search because it avoids getting stuck in this region of the search space.

3.5   Perturbation Operator

In (μ, λ) EAs, the perturbation operator largely determines the search capabilities of the algorithm. The mechanism proposed is based on modifying the channels allocated to a number of transceivers. It first has to select the set of TRXs to be modified and, next, it chooses the new channels which will be allocated. The two proposed methods are as follows:

1. TRX Selection: At each operation, one single site is perturbed. The way of selecting the site is to choose first a TRX t and then the site considered is the one at which t is installed. Two strategies for choosing t have been used:
   (a) Binary Tournament: It uses the same information from the simulator as the last greedy operation in the initialization method (see Section 3.4). Given two randomly chosen TRXs, this strategy returns the "hardest to deal with", i.e., the one which is preferred to be updated first. With this configuration, the perturbation mainly promotes intensification.
   (b) Random: The transceiver is randomly chosen using a uniform distribution from the whole set of TRXs. This strategy enhances the diversification capabilities of the algorithm.
   Since λ offsprings have to be generated at each step of the algorithm, we have studied several configurations in which λ_1 perturbations use the first strategy while λ_2 use the second one, so that $\lambda_1 + \lambda_2 = \lambda$. This will allow us to test different diversification/intensification tradeoffs in the EA.
2. Frequency Selection: Let s be the site chosen in the previous step. Firstly, s is assigned a hopefully interference-free frequency planning with the same strategy used in the initialization method (Fig. 5). We have therefore avoided the strongest intra-site interferences. The next step aims at refining this frequency plan by reducing the interferences with the neighboring sites. The strategy proceeds iterating through all the TRXs installed in s. Again, these TRXs are ranked in decreasing order with the accurate information coming from the simulator. Finally, for each TRX t, if a frequency f, different from the currently assigned one, allows both to keep the intra-site interference-free assignment and to reduce the interference from the neighbors, then t is assigned f; otherwise, it does nothing. All the available frequencies for t are examined. A pseudocode of the method is included in Fig. 6. Note that this procedure guarantees that no interference will occur among the TRXs installed in a site.

4   Experiments

In this section we turn to present the experiments conducted to evaluate the proposed $(1, \lambda)$ EAs. We firstly give some details of the GSM network instance used. The experiments with different configurations of the evolutionary optimizer are presented and analyzed afterwards. We have made 30 independent runs of each experiment. The results included are the median, $\tilde{x}$, and interquartile range, IQR, as measures of location (or central tendency) and statistical

```
1:  trxs = TRXsFromSite(s);
2:  applySimpleGreedyHeuristic(s);
3:  trxs = rank(trxs);
4:  for t in trxs do
5:      f = chooseMinimumInterferenceFrequency(t,neighbors(s));
6:      assign(t,f);
7:  end for
```

Fig. 6. Pseudocode of frequency selection strategy

dispersion, respectively. Since we are dealing with stochastic algorithms and we do want to provide the results with statistical confidence, the following analysis has been performed in all this work. Firstly, a Kolmogorov-Smirnov test has been performed and it shows that all but one dataset follow a normal (gaussian) distribution. Therefore, the non-parametric Kruskal-Wallis test has been used. We have considered confidence level of 95% (i.e., significance level of 5% or p-value under 0.05), which means that the differences are unlikely to have occurred by chance with a probability of 95%.

4.1   GSM Instance Used

Here, we want to provide the reader with details on the AFP instance which is being solved. The GSM network used has 711 sectors with 2,612 TRXs installed. That is, the length of the individuals in the EA is 2,612. Each TRX has 18 available channels (from 134 to 151). Figure 7 displays the network topology. Each triangle represents a sectorized antenna in which operate several TRXs. As it can be seen, the instance presents clustered plus non-clustered zones where no classical hexagonal cell shapes exist (typically used in simplified models of the problem). Additional topological information indicates that, on average, each TRX has 25.08 first order neighbors and 96.60 second order neighbors, thus showing the high complexity of this AFP instance, in which the available spectrum is much smaller than the average number of neighbors. Indeed, only 18 channels can be allocated to TRXs with 25.08 potential first order neighbors. We also want to remark that this real network operates with advanced interference reduction technologies and it employs accurate interference information which has been actually measured at a cell-to-cell level (neither predictions nor distance-driven estimations are used).

4.2   Results

We have conducted different experiments with several configurations of the $(1,\lambda)$ EA. Firstly, we have used two different values for λ, $\lambda = 10$ and $\lambda = 20$. For each value, five different schemes for generating the λ individuals have been tested by using several combinations of the two strategies proposed for selecting the TRXs (Section 3.5). All the configurations of the $(1,\lambda)$ EA stop when

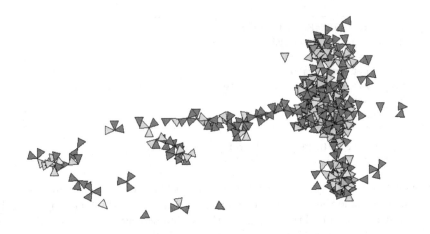

Fig. 7. Topology of the GSM instance used

100,000 function evaluations have been computed. Table 1 includes the obtained results, having the best one a grey colored background.

Let us start analyzing the results of our evolutionary algorithm. The configuration of the (1,10) EA in which the 10 offsprings are generated by using the Tournament strategy for selecting the TRXs ($\lambda_1 = 10$) has reached the best (lowest) cost: a value of 4680. The "+" symbol in the last row shows a successful Kruskal-Wallis test indicating that the differences are statistically significant. This configuration has the stressed intensification capabilities out of the 10 ones proposed due to two facts. On the one hand, since all of them run until 100,000 function evaluations have been computed, (1, 10) EA performs 10,000 iterations while the (1, 20) EA executes 5,000. Indeed, iterating for longer time means a larger exploitation of the search experience. On the other hand, the perturbation also uses the configuration with the more accentuated search intensification features: the Tournament selection strategy for the ten offsprings, in which those TRXs incurring in more problems (capacity overload, interferences, etc.) are chosen first. So, when the $(1, \lambda)$ EAs face the real-world AFP problem, they are able to profit from search intensification strategies which allow more accurate frequency plans to be computed. Note that, since both the traffic data and network configuration remain the same, a cost reduction means that fewer interferences exist in the network, i.e., several co-channels and adj-channel interferences have vanished. In this real-world AFP context, reducing this cost function leads the GSM operator to two possible scenarios. If the traffic demand is kept (or it is unusually decreased), the QoS for subscribers becomes higher. Otherwise, if this demand is increased and additional TRXs must be added, the network is ready to address its expansion also getting an acceptable QoS. The latter is specially important for the operators because it results in higher income for them.

Two additional facts can be deduced from Table 1. Firstly, all configurations of the (1, 10) EA improve upon any config where $\lambda = 20$. Averaging over all their configurations, (1,10) EAs get a cost of 4696.4, whereas (1,20) EAs obtain

Table 1. AFP costs when using ten different configurations of the $(1, \lambda)$ EA

Algorithm	Config λ_1	λ_2	Costs $\tilde{x}$	IQR
	0	10	4701	88
	3	7	4688	91
(1,10) EA	5	5	4708	104
	7	3	4705	113
	10	0	4680	148
	0	20	4725	141
	5	15	4741	100
(1,20) EA	10	10	4763	138
	15	5	4743	115
	20	0	4787	151
Kruskal-Wallis test			+	

a value 4751.8, which is a 1.17% worse. As it is explained before, this is because of the enhanced intensification capabilities of the $\lambda = 10$ setting. Secondly, the low IQR in all the cases indicates that all the proposals are very robust, reaching very accurate frequency plans in all the independent runs. Finally, we also want to mention that the execution times of all the $(1, \lambda)$ EAs are around 9,400 seconds on a Pentium IV, 2.4 GHz, and 512 MB of RAM. This value is the same for all the algorithms since all of them compute the same number of function evaluation and it implies a large effort to perform all these experiments (more than 32 days of computation).

5   Conclusions and Future Work

This paper describes the utilization of $(1, \lambda)$ EAs to solve the AFP problem in a real-world GSM network composed of 2,612 transceivers. Instead of using a mathematical formulation of this optimization problem, we have used a commercial application which allows the target frequency plannings to be accurately evaluated in a real scenario where current technologies are in use (e.g. frequency hopping, discontinuous transmission, dynamic power control, etc.).

We have implemented advanced operators for initializing the individuals and for generating the offspring. Then, different configurations for the evolutionary algorithms have been proposed and evaluated ($\lambda = 10$, $\lambda = 20$, different settings of the perturbation operator). The results show that those configurations promoting intensification have reached the best (lowest) costs, i.e., they have computed the frequency plans with the lower interference in the network.

As future work, we plan to develop new search operators and new metaheuristic algorithms to solve this problem. Their evaluation with the current instance and other real-world GSM networks is also an ongoing research line. The formulation of the AFP problem as a multiobjective optimization problem will be investigated as well.

References

1. Mouly, M., Paulet, M.B.: The GSM System for Mobile Communications. Mouly et Paulet, Palaiseau (1992)
2. Rapeli, J.: UMTS: Targets, system concept, and standardization in a global framework. IEEE Personal Communications **2** (1995) 30–37
3. Granbohm, H., Wiklund, J.: GPRS – general packet radio service. Ericsson Review (1999)
4. Furuskar, A., Naslund, J., Olofsson, H.: EDGE – enhanced data rates for GSM and TDMA/136 evolution. Ericsson Review (1999)
5. Aardal, K.I., van Hoesen, S.P.M., Koster, A.M.C.A., Mannino, C., Sassano, A.: Models and solution techniques for frequency assignment problems. 4OR **1** (2003) 261–317
6. FAP Web: (http://fap.zib.de/)
7. Kotrotsos, S., Kotsakis, G., Demestichas, P., Tzifa, E., Demesticha, V., Anagnostou, M.: Formulation and computationally efficient algorithms for an interference-oriented version of the frequency assignment problem. Wireless Personal Communications **18** (2001) 289–317
8. Eisenblätter, A.: Frequency Assignment in GSM Networks: Models, Heuristics, and Lower Bounds. PhD thesis, Technische Universität Berlin (2001)
9. Hale, W.K.: Frequency assignment: Theory and applications. Proceedings of the IEEE **68** (1980) 1497–1514
10. Blum, C., Roli, A.: Metaheuristics in Combinatorial Optimization: Overview and Conceptual Comparison. ACM Computing Surveys **35** (2003) 268–308
11. Glover, F.W., Kochenberger, G.A.: Handbook of Metaheuristics. Kluwer (2003)
12. Bäck, T.: Evolutionary Algorithms: Theory and Practice. Oxford University Press, New York, USA (1996)
13. Dorne, R., Hao, J.K.: An evolutionary approach for frequency assignment in cellular radio networks. In: Proc. of the IEEE Int. Conf. on Evolutionary Computation. (1995) 539–544
14. Smith, D.H., Allen, S.M., Hurley, S.: Characteristics of good meta-heuristics algorithms for the frequency assignment problem. Annals of Operations Research **107** (2001) 285–301
15. Schwefel, H.P.: Numerical Optimization of Computer Models. Wiley (1981)
16. Mishra, A.R.: Radio Network Planning and Optimisation. In: Fundamentals of Cellular Network Planning and Optimisation: 2G/2.5G/3G... Evolution to 4G. Wiley (2004) 21–54
17. Kampstra, P., van der Mei, R.D., Eiben, A.E.: Evolutionary computing in telecommunication network design: A survey. In Revision (2006)
18. Vidyarthi, G., Ngom, A., Stojmenović, I.: A hybrid channel assignment approach using an efficient evolutionary strategy in wireless mobile networks. IEEE Transactions on Vehicular Technology **54** (2005) 1887–1895

A New Metaheuristic for the Vehicle Routing Problem with Split Demands

Enrique Mota, Vicente Campos, and Ángel Corberán

Dep. Estasdística i Investigació Operativa. Universitat de València. Spain
mota@uv.es

Abstract. In this paper we present a metaheuristic procedure constructed for the special case of the Vehicle Routing Problem in which the demands of the clients can be split, i.e., any client can be serviced by more than one vehicle. The proposed algorithm, based on the scatter search methodology, produces a feasible solution using the minimum number of vehicles. The results obtained compare with the best results known up to date on a set of instances previously published in the literature.

1 Introduction

In this paper we consider a variant of the Vehicle Routing Problem (VRP) in which the demand of any client can be serviced by more than one vehicle, the Split Delivery Vehicle Routing Problem (SDVRP). This relaxation of the classical VRP was first proposed by Dror and Trudeau [8] and [9], who showed that important savings on the total solution cost could be obtained as well as a reduction in the total number of vehicles used in the solution. They also showed that this problem is also NP-hard.

Dror, Laporte and Trudeau [7] proposed a branch and bound algorithm for its exact resolution. A polyhedral study and a lower bound for the SDVRP were presented in [5], while some applications can be found in [15] and [16]. A worst-case study was conducted by Archetti, Savelsbergh and Speranza [2] as well as an evaluation of the situations when it pays to split the demands [3]. A tabu search procedure was developed by Archetti, Hertz and Speranza [1] and more recent heuristic procedures have been proposed by Wasil, Golden and Chen [17] and Archetti, Savelsbergh and Speranza [4].

The Vehicle Routing Problem with Split Demands is defined on an undirected and complete graph $G = (V,E)$, where $V = \{0,1,2,...n\}$ is the set of vertices (vertex 0 denotes the depot and $1,...,n$ represent the set of clients). Each edge $e = (i,j)$ has an associated cost or distance c_e between clients i and j. Moreover, each vertex has a known demand d_i ($d_0=0$) and there is a fleet of identical vehicles of capacity Q located at the depot. The objective is to find a set of routes, each one beginning and ending at the depot, such that:

- The demand of every client is satisfied,
- The sum of the demands serviced by any vehicle does not exceed its capacity Q, and
- The total cost, i.e. the sum of the costs of every edge in every route, is minimized.

C. Cotta and J. van Hemert (Eds.): EvoCOP 2007, LNCS 4446, pp. 121–129, 2007.

Other common variants of the VRP consider a servicing cost s_i for each client and an upper bound on the total distance traveled by any vehicle. The SDVRP with time windows has been studied in [10], where also special graphs have been considered.

We consider in this paper just the SDVRP version defined above, which is a very difficult problem but presents an outstanding characteristic, that makes it different from the classical VRP: *there is always a feasible solution using the minimum number of vehicles k*. It is easy to see that this minimum number corresponds to the smallest integer greater than or equal to $\dfrac{\sum_i d_i}{Q}$. This is not always true if the demand of a client can not be split, since in this case there is a Bin Packing Problem involved. To the explicit objective of minimizing the total solution cost, we add the implicit one of minimizing the number of vehicles used in the solution. A term in the objective function penalizing the excess of vehicles could be added, or bicriteria techniques could be taken into account, since it is possible in some instances to decrease the total cost by increasing the number of vehicles. Instead, we propose a Scatter Search procedure following the framework presented in [12] and [13], that generates a population of feasible solutions with the minimum number of routes (vehicles).

The paper is organized as follows: in Section 2 we describe the main features of the proposed metaheuristic and in Section 3 we present the computational results. Conclusions and future work are summarized in Section 4.

2 A Scatter Search Procedure

In this section we describe the main features of a Scatter Search procedure designed for the SDVRP. This is, as far as we know, the first time that such a technique is applied to this routing problem.

2.1 Creating a Population

We have adapted two standard VRP heuristic procedures to the split demands case in order to obtain SDVRP feasible solutions. The first one, called here Big Tour, uses the Lin and Kernighan [14] heuristic to build a giant tour through the n clients and the depot. From this tour it is always possible to obtain k routes and, thus, a feasible SDVRP solution: Clients are grouped into routes following the tour and the last client in a given route is the one that usually produces an overload on the vehicle; in this case the client's demand is split so that the total demand serviced by the vehicle equals Q and a new vehicle leaves the depot, visits this client again and follows the tour.

We can take into account the difference between the total capacity kQ and the total demand and adjust the load in each vehicle so that the solution finally obtained uses k balanced (in terms of load) routes. In this way, we avoid obtaining a solution having $k-1$ routes with load Q and a last route with usually a very small load.

The same Big Tour is used to generate additional solutions, all of them following the same sequence of clients but starting each one at a different client. In order to obtain solutions that differ substantially, the starting clients are selected in a nonconsecutive order.

The procedure Big Tour is designed to generate half of the population of feasible solutions, of size P.

The second procedure is a modified and accelerated version of the classical Clarke and Wright parallel savings algorithm [6]. According to this procedure, from an initial solution consisting of n return trips to each client, the best available saving, computed as $s_{ij} = c_{0i} + c_{0j} - \lambda c_{ij}$, is used to merge the single routes $(0,i,0)$ and $(0,j,0)$ into a new route $(0,i,j,0)$ and the procedure is repeated until no merge is feasible, in terms of vehicle capacity, or there are no more available savings. For each client, its neighborhood is computed as the subset of its closest clients, and only these savings are calculated. We allow to split the demand of a client l only when the best available saving corresponds to merging a given route r with a return trip from client l and the total demand exceeds the vehicle capacity Q; in this case, part of the demand of client l is serviced in route r and we maintain a return trip from client l with the unsatisfied demand. The procedure does not guarantee a feasible solution using the minimum number of vehicles but in all our computational experiences, feasible solutions using k vehicles are obtained. In order to generate more than one solution, we prohibit half of the savings used in a solution when computing the next one. Savings are prohibited with probabilities directly proportional to the frequency of use of each saving in the previously generated solutions. This procedure generates half of the population of feasible solutions.

2.2 Improving a Feasible Solution

Local search is applied to each solution in the original population in order to reduce its cost, if possible. We have implemented procedures for client moves, such as the *1-1 interchanges*, consisting of interchanging one client from a route with another client in another route, and *1-0 exchanges*, consisting of shifting one client from one route to another route. These moves are applied to every non split client. We have also implemented *two-split changes*, that take a client out from every route visiting it and look for a pair of routes that, jointly, could service its demand. When such improvements are no longer possible, the routes in the solution are re-optimized using a 2-opt procedure or the more complex Lin and Kernighan algorithm. The same procedures are applied to a feasible solution entering the reference set, as described in the next subsection.

2.3 The Reference Set

The P feasible solutions in the population are ordered according to the cost and b of them are selected to be in the reference set. One half corresponds to the best feasible solutions and the remaining solutions add the necessary diversity to this set, since they correspond to those solutions in the population that are the most different when compared to the best ones. As a measure of the difference between two solutions we compute the total number of edges in one solution but not in the other. Each pair of solutions in the reference set is combined to produce another solution, that enters the set only when its cost is greater that the cost of the worst solution, that is eliminated. The overall procedure stops when, after every possible combination (one iteration), no new feasible solution enters in the reference set.

2.4 The Combination Method

We have devised a procedure that captures the essential characteristics of a feasible SDVRP solution and tries to maintain those that could be satisfied by the good solutions. In order to do that, for each solution in the reference set we define a set of *critical* clients, consisting of:

1. all its split clients,
2. all the clients (if distinct) in routes with just 1 or 2 clients,
3. the client (if distinct) whose removal from a route produces the greatest saving cost, for each route with at least 3 clients, and finally
4. each client (if distinct) such that at least one among its three closest neighbours belongs to a different route.

When combining feasible solutions A and B in the reference set (note that combining solutions B and A is also possible and produces a different combination) we consider, in turn, a *critical client* in A, in classes 1 to 3 above, and we apply the recommendation for this client in solution B. If it is a split client in B, we consider that there is no recommendation and so we take the next *critical client* in A; otherwise we consider its two adjacent clients, say α and β, and we route in the combined solution the *critical client* in the best position, either after client α or just before client β. Once the insertion is made, we declare tabu any move involving the *critical client* in the combined solution.

When all the *critical clients* of A have been considered, the combination method has produced a new and maybe unfeasible solution because of the load in each route. A routine is then applied that considers some non tabu moves aimed at obtaining a feasible solution.

Each time a feasible solution is obtained as a combination of two solutions in the reference set, the improve procedures described in subsection 2.2 are applied. Once all the possible comparisons have been considered, if no new solution enters the reference set, we augment the set of *critical clients* of each solution in the reference set by including those in case 4 above. Then, all the possible combinations are again considered. The new neighborhood created for each solution replaces the usual rebuilding phase.

3 Computational Experiments

We present the results obtained by the Scatter Search procedure on a set of instances following the proposal made by Dror and Trudeau. We have repeated the generation parameters used by Archetti et al, so that the results can be compared.

3.1 The Instances

In order to test our algorithm, we have considered the same set of instances used by Archetti, Hertz and Speranza [1]. They generated the instances starting from the VRP problems 1 to 5, 11 and 12 taken from [11], that have between 50 and 199 customers, and computed the demands of the customers in the following way. First, two parameters α and γ ($\alpha \leq \gamma$) are chosen in the interval [0,1]. Then, the demand d_i of customer i is set equal to $d_i = \lfloor \alpha\, Q + \delta\, (\gamma - \alpha)\, Q \rfloor$, where δ is a random number in [0,1]. As in [8], Archetti, Hertz and Speranza have considered the following combinations of parameters (α, γ): (0.01, 0.1), (0.1, 0.3), (0.1, 0.5), (0.1, 0.9), (0.3, 0.7) and (0.7, 0.9). The procedure is equivalent to generate the demands of an instance in the interval $(\alpha Q, \gamma Q)$. Considering the case where the original demands are not changed, a total of 49 instances is obtained.

3.2 Computational Results

Computational results are summarized in Table 1, which compares the results obtained by the Scatter Search procedure with those obtained with two Tabu Search (TS) algorithms presented in [1]. Comparisons with the results presented in [4] and [17] were not possible since the instances and other significant details were not available to the authors. Other characteristics of the feasible solutions are also included in the table. The first two columns show the instance name, which also indicates the number of clients, and the corresponding interval where demands have been generated. Instances in the first seven rows have original demands. Column 3 presents the best value (z) obtained by our procedure SS; a value in bold indicates that this value is at least as good as all the ten values obtained by applying the SPLITABU or the SPLITABU-DT procedures ([1]). All the values have been obtained using as distance between clients i and j the cost c_{ij}:

$$c_{ij} = round\left(10000\sqrt{(x_i - x_j)^2 + (y_i - y_j)^2}\right) \qquad (1)$$

Column 4 indicates the number of vehicles in the feasible solution obtained (k), which corresponds always to the minimum number. Total time in seconds is presented in column 5 (T). The procedure was implemented in C and run on a PC Pentium IV, 1Gb Ram, CPU 2.40 GHz.

The minimum solution value (zmin) among the five executions that each instance is run with SPLITABU and SPLITABU-DT is shown in columns 6 and 9, respectively. Similarly, columns 7 and 10 present the average solution value for the 5 runs (zmean). Columns 8 and 11 give the average times, in seconds of a PC Pentium IV, 256 Mb Ram, CPU 2.40 GHz. Finally, column 12 gives the number of vehicles in the feasible solution as presented in [2].

Considering the original demands, the quality of the solutions is similar and every solution uses the minimum number of vehicles. Note that the values obtained by the Scatter Search algorithm were produced maintaining all the parameters unchanged for all the instances and only one execution per instance, in order to make a fair comparison. The number of vehicles used is not available for both TS methods on the second group of instances and the solutions obtained by the SS are slightly better than the best solutions obtained with the TS procedures. When the demand is generated in the interval $(0.1Q, 0.3Q)$ the solution's quality is similar but the number of vehicles in the solutions obtained with the Tabu Search procedures is no longer the minimum one and the difference reaches 3 vehicles in 2 out of 6 instances. On the remaining instances, the SS solutions are worse than the ones obtained with the TS procedures. However, this could be explained by the fact that the SS algorithm is designed to find solutions with the minimum number of vehicles while the TS algorithms minimize the

Table 1. Computational results

prob	demand	SS			SP-TABU			SP-TABU-DT			
		z	k	T	zmin	zmean	T	zmin	zmean	T	k
p1-50		5310248	5	24.80	5276751	5300570	17	5306520	5335535	13	5
p2-75		**8397536**	10	61.66	8404529	8516729	64	8458221	8495410	36	10
p3-100		8358235	8	108.80	8341357	8461844	60	8333566	8356191	58	8
p4-150		**10569181**	12	261.28	10570279	10621988	440	10643847	10698369	389	12
p5-199		13404449	16	352.31	13573455	13678177	1900	13399777	13428515	386	16
p6-120		**10429739**	7	131.34	10763753	10847331	40	10535425	10560148	38	7
p7-100		8209220	10	108.41	8195581	8226045	86	8195581	8253184	49	
p1-50	0.01-0.1	**4607896**	3	26.86	4607896	4638532	9	4607896	4637571	5	
p2-75	0.01-0.1	6026681	4	68.80	6041186	6076579	42	6016174	6052376	13	
p3-100	0.01-0.1	**7296712**	5	125.06	7572459	7727881	59	7374514	7522012	31	
p4-150	0.01-0.1	8830484	8	352.09	8864415	8949843	258	8819995	8909533	173	
p5-199	0.01-0.1	**10395090**	10	963.84	10452620	10735985	754	10478739	10562679	526	
p6-120	0.01-0.1	**9795739**	6	163.28	10641948	10957537	61	10760897	10846959	42	
p7-100	0.01-0.1	**6338034**	5	80.56	6407126	6628037	71	6358865	6487359	58	
p1-50	0.1-0.3	7695976	10	26.31	7507084	7644041	27	7515968	7614021	22	11
p2-75	0.1-0.3	**10740110**	15	86.02	10835052	10990297	78	10879305	10953225	45	16
p3-100	0.1-0.3	**14164821**	20	98.00	14189700	14288683	122	14193762	14248114	96	22
p4-150	0.1-0.3	19747047	29	10.06	19284742	19406720	545	19079235	19182459	393	32
p5-199	0.1-0.3	24350759	38	19.11	24120235	24199773	1224	23780537	23841545	755	41
p6-120	0.1-0.3	**27831006**	23	11.33	28520739	29009898	516	29145714	29187092	143	26
p7-100	0.1-0.3	14234869	20	151.25	14531928	14709592	85	14379543	14620077	146	
p1-50	0.1-0.5	10259080	15	3.84	9940561	10076838	56	9972128	10086663	28	16
p2-75	0.1-0.5	14846153	22	6.09	14407823	14501086	71	14321606	14436243	123	24
p3-100	0.1-0.5	19261522	29	7.55	18788510	18878331	206	18857414	18947210	136	33

Table 1. (*continued*)

prob	demand	SS			SP-TABU			SP-TABU-DT			
		z	k	T	zmin	zmean	T	zmin	zmean	T	k
p4-150	0.1-0.5	26499703	43	16.17	26168532	26340901	564	26089113	26327126	739	49
p5-199	0.1-0.5	33107094	56	20.64	32765665	32981871	3811	32473125	32844723	2668	63
p6-120	0.1-0.5	**39962916**	34	63.80	41231758	41667801	259	41311259	42061210	268	40
p7-100	0.1-0.5	20223019	29	41.23	19924980	20300366	188	19815453	20299948	293	
p1-50	0.1-0.9	15807670	25	3.91	14815710	14939233	34	14438367	14699221	61	26
p2-75	0.1-0.9	22330822	37	6.64	21131636	21212778	311	21072124	21244269	193	41
p3-100	0.1-0.9	29323441	48	9.16	28116117	28266122	412	27467515	27940774	649	56
p4-150	0.1-0.9	41856832	73	25.03	39785729	40062807	1822	38497320	39097249	2278	84
p5-199	0.1-0.9	50856358	93	71.09	50058905	50396524	2598	47374671	48538254	3297	107
p6-120	0.1-0.9	63614574	56	15.86	65072461	65603347	1037	62596720	65839735	878	67
p7-100	0.1-0.9	31874383	48	9.08	31192745	31580865	523	30105041	31015273	260	
p1-50	0.3-0.7	15680385	25	4.25	14941462	15093118	52	14870198	14969009	49	26
p2-75	0.3-0.7	22288975	37	7.14	21666219	21759879	184	21497382	21605050	129	39
p3-100	0.3-0.7	29863298	49	10.36	28958054	29147859	454	27642538	28704954	810	53
p4-150	0.3-0.7	41856832	73	19.38	41228032	41467544	1512	39671062	40396994	3008	80
p5-199	0.3-0.7	52650121	96	120.28	53329707	53689192	2279	50014512	51028379	3566	103
p6-120	0.3-0.7	64810943	58	17.16	67207650	68663390	477	64330110	66395522	659	65
p7-100	0.3-0.7	32487607	49	9.73	31631372	32220334	411	28821235	30380225	778	
p1-50	0.7-0.9	23124751	40	4.13	21733326	21763923	160	21483778	21652085	106	42
p2-75	0.7-0.9	33878605	60	7.66	32184116	32294627	437	31381780	31806415	869	61
p3-100	0.7-0.9	45809798	80	12.06	43619465	43687723	1891	42788332	43023114	1398	82
p4-150	0.7-0.9	64794550	119	131.91	63345083	63542058	8783	60998678	61963577	10223	123
p5-199	0.7-0.9	83237230	158	165.28	82071543	83439543	11347	76761141	79446339	21849	162
p6-120	0.7-0.9	101583160	95	20.17	103067404	105505745	2033	100726022	103040778	1826	99
p7-100	0.7-0.9	50652558	80	9.19	49334893	49607513	1865	47735921	48677857	1004	

total distance traveled and note that the solutions obtained with the TS methods always use a bigger number of vehicles that, in some cases increases up to 14 vehicles as in instance p5-199 with demands in $(0.1Q, 0.9Q)$.

4 Conclusions and Further Research

The first results obtained with the Scatter Search procedure indicate that it is able to obtain good feasible solutions within a reasonable computing time. When the demands are well over half the capacity of the vehicle the values of the solutions are not so good, but note that we only consider solutions with the minimum number of vehicles and we think that this fact compensates having solutions with longer routes. The set of published and available instances is limited and quite small. In the future, we want to work on the elaboration of bigger test instances that will be available and include some other refinements to the procedure. Among them, another generator of feasible solutions and more procedures to be applied to the unfeasible solutions produced in the combination phase.

Acknowledgments. The authors want to acknowledge the support of the Spanish Ministerio de Educación y Ciencia, through grants MTM 2006-14961-C05-02 and TIN2006-02696. We are also grateful to C. Archetti and M.G. Speranza that kindly facilitated the instances used in the computational experiences.

References

1. Archetti, C., Hertz, A., Speranza, M.G.: A tabu search algorithm for the split delivery vehicle routing problem. Transportation Science, 39 (2005) 182 -187
2. Archetti, C., Savelsbergh, M.W.P., Speranza, M.G.: Worst-case analysis for split delivery vehicle routing problems. To appear in Transportation Science
3. Archetti, C., Savelsbergh, M.W.P., Speranza, M.G.: To Split or Not to Split: That is the Question. To appear in Transportation Research
4. Archetti, C., Savelsbergh, M.W.P., Speranza, M.G.: An Optimization-Based Heuristic for the Split Delivery Vehicle Routing Problem. Working paper.
5. Belenguer, J.M., Martínez, M.C., Mota, E.: A lower bound for the split delivery vehicle routing problem. Operations Research 48 (2000) 801-810
6. Clarke, G., Wright, J.V.: Scheduling of vehicles from a central depot to a number of delivery points. Operations Research 12 (1964) 568-581
7. Dror, M.,Laporte, G., Trudeau, P.: Vehicle Routing with Split Deliveries. Discrete Applied Mathematics 50 (1994) 239-254
8. Dror, M., Trudeau, P.: Savings by split delivery routing. Transportation Science 23 (1989) 141-145
9. Dror, M., Trudeau, P.: Split Delivery Routing. Naval Research Logistics 37 (1990) 383-402
10. Frizzell, P.W., Giffin, J.W.: The split delivery vehicle routing problem with time windows and grid network distances. Computers and Operations Research 22 (1995) 655-667.
11. Gendreau, M., Hertz, A., Laporte, G.: A tabu search heuristic for the vehicle routing problem. Management Science 40 (1994) 1276–1290.

12. Glover, F.: A template for scatter search and path relinking. In Hao, J.-K., Lutton, E., Schoenauer, M., Snyers, D. (eds.), Artificial Evolution. Lecture Notes in Computer Science 1363 (1998) 13-54
13. Laguna, M., Martí, R.: Scatter Search – Methodology and implementations in C. Kluwer Academic Publishers, Boston (2003)
14. Lin, S., Kernighan, B.W.: An effective heuristic algorithm for the travelling salesman problem. Operations Research 21 (1973) 498-516
15. Mullaseril, P.A., Dror, M., Leung, J.: Split-delivery routing in livestock feed distribution. Journal of the Operational Research Society 48 (1997) 107-116
16. Sierksma, G., Tijssen, G.A.: Routing helicopters for crew exchanges on off-shore locations. Annals of Operations Research 76 (1998) 261-286
17. Wasil, E., Golden, B., Chen, S.: The Split Delivery Vehicle Routing Problem: Applications, algorithms, test problems and computational results. Presented at EURO 2006. Iceland

Generation of Tree Decompositions by Iterated Local Search

Nysret Musliu

Vienna University of Technology, Karlsplatz 13, 1040 Vienna, Austria

Abstract. Many instances of NP-hard problems can be solved efficiently if the treewidth of their corresponding graph is small. Finding the optimal tree decompositions is an NP-hard problem and different algorithms have been proposed in the literature for generation of tree decompositions of small width. In this paper we propose a novel iterated local search algorithm to find good upper bounds for treewidth of an undirected graph. We propose two heuristics, and their combination for generation of the solutions in the construction phase. The iterated local search algorithm further includes the mechanism for perturbation of solution, and the mechanism for accepting solutions for the next iteration. The proposed algorithm iteratively applies the heuristic for finding good elimination ordering, the acceptance criteria, and the perturbation of solution. We proposed and evaluated different perturbation mechanisms and acceptance criteria. The proposed algorithms are tested on DIMACS instances for vertex coloring, and they are compared with the existing approaches in literature. Our algorithms have a good time performance and for 17 instances improve the best existing upper bounds for the treewidth.

1 Introduction

The concept of tree decompositions is very important due to the fact that many instances of constraint satisfaction problems and in general NP-hard problems can be solved in polynomial time if their treewidth is bounded by a constant. The process of solving problems with bounded treewidth includes two phases. In the first phase the tree decompositions with small upper bound for treewidth are generated. The second phase includes solving a problem (based on the generated tree decomposition) with a particular algorithm such as for example dynamic programming. The efficiency of solving of problem based on its tree decompositions depends from the width of tree decompositions. Thus it is of high interest to generate tree decompositions with small width.

In this paper we investigate the generation of tree decompositions of undirected graphs. The concept of tree decompositions has been first introduced by Robertson and Seymour ([11]):

Definition 1. *(see [11], [9]) Let $G = (V, E)$ be a graph. A tree decomposition of G is a pair (T, χ), where $T = (I, F)$ is a tree with node set I and edge set F, and $\chi = \{\chi_i : i \in I\}$ is a family of subsets of V, one for each node of T, such that*

C. Cotta and J. van Hemert (Eds.): EvoCOP 2007, LNCS 4446, pp. 130–141, 2007.

1. $\bigcup_{i \in I} \chi_i = V$,
2. *for every edge* $(v, w) \in E$, *there is an* $i \in I$ *with* $v \in \chi_i$ *and* $w \in \chi_i$, *and*
3. *for all* $i, j, k \in I$, *if* j *is on the path from* i *to* k *in* T, *then* $\chi_i \cap \chi_k \subseteq \chi_j$.

The width of a tree decomposition is $max_{i \in I} |\chi_i| - 1$. The treewidth of a graph G, denoted by $tw(G)$, is the minimum width over all possible tree decompositions of G.

Figure 1 shows a graph G (19 vertices) and a possible tree decomposition of G. The width of shown tree decomposition is 5.

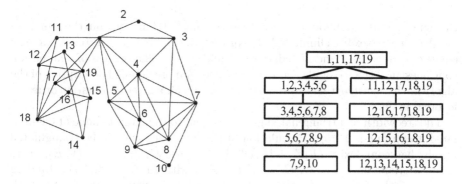

Fig. 1. A graph G (left) and a tree decomposition of G (right)

For the given graph G the treewidth can be found from its triangulation. Further we will give basic definitions, explain how the triangulation of graph can be constructed, and give lemmas which give relation between the treewidth and the triangulated graph.

Two vertices u and v of graph $G(V, E)$ are neighbours, if they are connected with an edge $e \in E$. The neighbourhood of vertex v is defined as: $N(v) := \{w | w \in V, (v, w) \in E\}$. A set of vertices is clique if they are fully connected. An edge connecting two non-adjacent vertices in the cycle is called chord. The graph is triangulated if there exist a chord in every cycle of length larger than 3.

A vertex of a graph is simplicial if its neighbours form a clique. An ordering of nodes $\sigma(1, 2, \ldots, n)$ of V is called a perfect elimination ordering for G if for any $i \in \{1, 2, \ldots, n\}$, $\sigma(i)$ is a simplicial vertex in $G[\sigma(i), \ldots, \sigma(n)]$ [3]. In [4] it is proved that the graph G is triangulated if and only if it has a perfect elimination ordering. Given an elimination ordering of nodes the triangulation H of graph G can be constructed as following. Initially $H = G$, then in the process of elimination of vertices, the next vertex in order to be eliminated is made simplicial vertex by adding of new edges to connect all its neighbours in current G and H. The vertex is then eliminated from G. The process of elimination of nodes from the given graph G is illustrated in Figure 2. Suppose that we have given the following elimination ordering: $10, 9, 8, \ldots$. The vertex 10 is first eliminated from G. When this vertex is eliminated no new edges are added in the graph G and H (graph H is not shown in

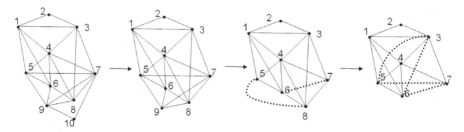

Fig. 2. Illustration of the elimination of nodes 10, 9, and 8 in process of constructing of triangulated graph

the figure), as all neighbours of node 10 are connected. Further from the remained graph G the vertex 9 is eliminated. To connect all neighbours of vertex 9, two new edges are added in G and H (edges $(5,8)$ and $(6,7)$). The process of elimination continues until the triangulation H is obtained. A more detailed description of the algorithm for constructing a graph's triangulation for a given elimination ordering is found in [9].

The treewidth of a triangulated graph is equal to the largest clique of triangulated graph minus 1 ([5]). Calculation of the largest clique for the triangulated graphs has complexity $O(|V| + |E|)$ ([5]). For every graph $G = (V, E)$, there exists a triangulation of G, $\overline{G} = (V, E \bigcup E_t)$, with $tw(\overline{G}) = tw(G)$. Finding the treewidth of a graph G is equivalent to finding a triangulation $\overline{G}$ of G with minimum clique size. The last two lemmas can be found in [9].

1.1   Algorithms for Tree Decompositions

For the given graph and integer k, deciding whether the graph has a tree decomposition with a treewidth at most k is an NP-hard problem [1]. To solve this problem different complete and heuristic algorithms have been proposed in the literature. Examples of complete algorithms for tree decompositions are [12] and [6]. Gogate and Dechter [6] reported good results for tree decompositions by using the branch and bound algorithm. They showed that their algorithm is superior compared to the algorithm proposed in [12]. The branch and bound algorithm proposed in [6] applies different pruning techniques, and provides anytime solutions, which are good upper bounds for tree decompositions.

Heuristic techniques for generation of tree decompositions with small width are mainly based on searching for a good perfect elimination ordering of graph nodes. Several heuristics that run in polynomial time have been proposed for finding a good elimination ordering of nodes. These heuristics select the ordering of nodes based on different criteria, such as the degree of the nodes, the number of edges to be added to make the node simplicial (the node which neighbours are fully connected), etc. Maximum Cardinality Search (MCS) proposed by Tarjan and Yannakakis ([13] constructs the ordering of nodes iteratively by picking the next node which has the largest number of neighbors in the ordering (the ties are broken randomly). The min-fill heuristics picks iteratively the node which adds the smallest number of edges when eliminated. Min-degree heuristic picks the next vertex to be eliminated based on its degree. The next node to be eliminated is chosen based on the smallest degree. According to [6] the min-fill heuristic performs better than MCS and min-degree heuristic. Min-degree heuristic has been improved by Clautiaux et al ([3] by adding a new criterion based on the lower bound of the treewidth for the graph obtained when the node is eliminated. For other types of heuristics based on the elimination ordering of nodes see [9].

Metaheuristic approaches have also been used for tree decompositions. Simulated annealing was used by Kjaerulff ([8]). Application of genetic algorithm for tree decompositions is presented in [10]. A tabu search approach for generation of the tree decompositions has been proposed by Clautiaux et al [3]. The authors reported good results for DIMACS vertex coloring instances ([7]). Their approach improved the previous results in literature for 53% of instances. Some of the results in [3] have been further improved by Gogate and Dechter [6]. The reader is referred to [2] for other approximation algorithms, and the information for lower bounds algorithms.

In this paper we propose new heuristic algorithms with the aim of improving existing upper bounds for tree decompositions and reducing the running time of algorithms for different problems. Two simple heuristics for searching in the elimination ordering of nodes are proposed. These local heuristics are based on changing of positions of nodes in ordering, which cause the largest clique when eliminated. The proposed heuristics are exploited by a new iterated local search algorithm in the construction phase. The proposed iterative local search algorithm applies iteratively the construction heuristic and additionally includes the perturbation mechanism and the acceptance criteria. These algorithms have been applied in 62 DIMACS instances for vertex coloring. For several problems we report new upper bounds for the treewidth, and for most of problems the tree decomposition is generated in a reasonable amount of time. Our results have been compared with the results reported in [9],[6], and [3], which to our best knowledge report the best results known yet in literature considering the width of tree decompositions for these instances. For up to date information for the best upper and lower bounds for treewidth for different instances the reader is referred to TreewidthLIB:http://www.cs.uu.nl/ hansb/treewidthlib/.

2   An Iterative Local Search Algorithm

As described in the previous section, the generation of tree decomposition with small width can be done by finding an appropriate elimination ordering which produces a triangulated graph with smallest maximum clique size. In this section we present an algorithm which searches among the possible ordering of nodes to find a small treewidth for the given graph. The algorithm contains a local search heuristic for constructing a good ordering, and the iterative process, during which the algorithm calls the local search techniques with the initial solution that is produced in previous iteration. The algorithm includes also a mechanism for acceptance of a candidate solution for the next iteration. Although the constructing phase is very important, choosing the appropriate perturbation in each iteration as well as the mechanism for acceptance of solution are also very important to obtain good results using an iterative local search algorithm. The proposed algorithm is presented in Algorithm 1.

Algorithm 1. Iterative heuristic algorithm - IHA

Generate initial solution $S1$

while Number of Iterations < MAXIterations **do**
    Get solution $S2$ from the execution of one of local search techniques proposed in the next section. The local search technique uses the solution $S1$ as an initial solution

    if Solution $S2$ fulfils the acceptance criteria **then**
        $S1 = S2$
    end if

    Apply perturbation in solution $S1$

end while

As an initial solution we use an order of nodes as they appear in the input. Better initial solutions can also be constructed by using other heuristics which run in polynomial time, such as Maximum Cardinality Search, min-fill heuristic, etc. However, as the proposed method usually finds fast a solution produced by these heuristics, our algorithm starts with very poor initial solution.

2.1   Local Search Techniques

We propose two local search methods for generation of a good solution which will be used as an initial solution with some perturbation in the next call of the same local search algorithm. Both techniques are based on the idea of moving only those vertices in the ordering, which cause the largest clique during the elimination process. The motivation for using this method is the reduction of the number of solutions that should be evaluated. The first proposed technique (LS1) is presented in Algorithm 2.

As we see above, the proposed algorithm applies a very simple heuristic. A vertex is chosen randomly among the vertices that have the same number of neighbourhood vertices as the largest clique obtained during the elimination process. We experimented with two types of moves. In the first variant the vertex is inserted in a random position in the elimination ordering, while in the second variant the vertex is swapped with another vertex located in a randomly selected position, i.e. the two chosen vertices change their position in the elimination ordering. The heuristic will stop if the solution is not improved after a certain number of iterations. Although this is a very simple heuristic, using it alone does

---

Algorithm 2. Local Search Algorithm 1 - LS1 (InputSolution)

---

while NrNotImprovments < MAXNotImprovments **do**

Select a vertex in the elimination ordering which causes the largest clique (ties are broken randomly if there are several vertices which cause the cliques with the same size)

Swap this vertex with another vertex located in a randomly chosen position

end while

---

not produce good results for the tree decompositions. Whereas combination with the iterative method (see Algorithm 1) it generates good results.

The second proposed heuristic (LS2) is presented in Algorithm 3. This technique is similar to algorithm LS1. However, in this technique in each iteration we apply the same procedure as in the LS1 with some probability p, whereas with probability $1 - p$, the best solution is selected (ties are broken randomly) from the neighbourhood of solution. The neighbourhood of a solution is obtained by generation of all solutions which are obtained by swapping of selected vertex with all its neighbour vertices in the graph.

Algorithm 3. Local Search Algorithm 2 - LS2 (InputSolution)

---

while NrNotImprovments < MAXNotImprovments **do**

With probability p:

Select a vertex in the elimination ordering which causes the largest clique (ties are broken randomly)

Swap this vertex with another vertex located in the randomly chosen position

With probability $1 - p$:

Select a vertex in the elimination ordering which causes the largest clique (ties are broken randomly)

Generate neighbourhood of the solution by swapping the selected vertex with its neighbours, i.e. all solutions are generated by swapping the selected vertex with its neighbours

Select the best solution from the generated neighbourhood

end while

2.2 Perturbation

During the perturbation phase the solution obtained by local search procedure is perturbed and the newly obtained solution is used as an initial solution for the new call of the local search technique. The main idea is to avoid the random restart. Instead or random restart the solution is perturbed with a bigger move(s) as those applied in the local search technique. This enables some diversification that helps to escape from the local optimum, but avoids beginning from scratch (as in case of random restart), which is very time consuming. We propose three perturbation mechanisms for the solution:

- RandPert: N vertices are chosen randomly and they are moved into new random positions in the ordering.
- MaxCliquePer: All nodes that produce the maximal clique in the elimination ordering are inserted in a new randomly chosen positions in the ordering.
- DestroyPartPert: All nodes between two positions (selected randomly) in the ordering are inserted in the new randomly chosen positions in the ordering.

Determining the number of nodes N that will be moved is complex and may be dependent on the problem. To avoid this problem we propose an adaptive perturbation mechanism that takes into consideration the feedback from the search process. The number of nodes N varies from 2 to 10, and the algorithm begins with small perturbation with $N = 2$. If during the iterative process (for a determined number of iterations) the local search technique produces solutions with same tree width for more than 20% of cases, the size of perturbation is increased by 1, otherwise the size of N will be decreased by 1. This enables an automatic change of perturbation size based on the repetition of solutions with the same width. We applied each perturbation mechanism separately, and also considered combination of two perturbations, so that one perturbation is applied for the certain number and another perturbation is applied for the certain next number of iterations.

2.3   Acceptance of Solution in Iterated Algorithm

Different techniques can be applied for acceptance of the solution obtained by the local search technique. If the solution is accepted it will be perturbed and will serve as an initial solution for the next call of one of the local search techniques. We experimented with the following variants for acceptance of solution for the next iteration (see Algorithm 1):

- Solution S2 is accepted only if it has a better width than the solution S1.
- Solution S2 is always accepted.
- Solution S2 is accepted if its treewidth is not larger than the treewidth of the best yet found solution minus x, where x is an integer.

2.4   Setting of Parameters

Using our algorithm we experimented with two proposed local search techniques for construction phase, different perturbation, different acceptance criteria, swap neighbourhood, and different termination criteria for the local search procedures. For algorithm LS2 we experimented with probability $p = 10, 30, 50$. Considering the acceptance of solution in iterated local search we experimented with three variants described in Section 2.3. For the third variant we experimented with $x = 2$ and $x = 3$. We did experiments with three types of perturbations: RandPert, MaxCliquePer, and DestroyPartPer. Additionally, we experimented with combination of RandPert and MaxCliquePer. The current best results presented in this paper are obtained with the iterative heuristic algorithm (IHA) and these parameters: LS1 algorithm (see Algorithm 2) is used in the construction phase and this algorithm stops if the solution does not improve for 10 iterations ($MAXNotImprovments = 10$). In the perturbation phase are used both RandPert and MaxCliquePer perturbations. Initially RandPert with $N = 2 - 10$ is applied. Further the algorithm switches alternatively between two perturbations RandPert and MaxCliquePer, when IHA runs for 100 iterations without improvement of a solution. For accepting of solution in IHA the third variant is used. The solution produced in construction phase is accepted if its width is not more than the width of the best current solution plus 3.

3   Computational Results

In this section we report on computational results obtained with the current implementation of methods described in this paper. The results for 62 DIMACS vertex coloring instances are given. These instances have been used for testing of several methods for tree decompositions proposed in the literature (see [9], [3], and [6]). Our algorithms have been implemented in C++ and the current experiments were performed with a Intel Pentium 4 CPU 3GHz, 1GB RAM.

We compare our results with the results reported in [9], [3], and [6]. The results reported in [9] are obtained in Pentium 3.8GHz processor. Results reported in [3] are obtained with Pentium 3 1GHz processor, and the results reported in [6] are obtained with Pentium-4 2.4 Ghz, 2GB RAM machine. To our best knowledge these papers present the best existing upper bounds for treewidth for these 62 instances.

In Table 1 the results for the treewidth for DIMACS graph coloring instances are presented. First and second columns of the table present the instances and the number of nodes and edges for each instance. In column KBH are shown the best results obtained by algorithms in [9]. The TabuS column presents the results reported in [3], while the column BB shows the results obtained with the branch and bound algorithm proposed in [6]. The last two columns present results obtained by our algorithm proposed in this paper. In our algorithm are executed three runs for each instance. In column IHA-best is given the best width obtained in three runs for each instance, and the column IHA-AVG gives the average of treewidth over 3 runs.

In Table 2 for each instance is given the time (in seconds) needed to produce the treewidth presented in Table 1 for all algorithms. The time results given in [6] present the time in which the best solutions are found. The results given in [3] present the time of the overall run of the algorithm in one instance (number of iterations is 20000 and the algorithm stops after 10000 non-improving solutions). For our algorithm are given the time results for finding of best solutions

Table 1. Algorithms comparison regarding treewidth for DIMACS graph coloring instances

Instance	$|V|/|E|$	KBH	TabuS	BB	IHA-best	IHA-AVG
anna	138 / 986	12	12	12	12	12
david	87 / 812	13	13	13	13	13
huck	74 / 602	10	10	10	10	10
homer	561 / 3258	31	31	31	31	31
jean	80 / 508	9	9	9	9	9
games120	120 / 638	37	33	-	**32**	32
queen5_5	25 / 160	18	18	18	18	18
queen6_6	36 / 290	26	25	25	25	25
queen7_7	49 / 476	35	35	35	35	35
queen8_8	64 / 728	46	46	46	**45**	45.3
queen9_9	81 / 1056	59	58	59	58	58
queen10_10	100 / 1470	73	72	72	72	73
queen11_11	121 / 1980	89	88	89	88	88.7
queen12_12	144 / 2596	106	**104**	110	105	106.3
queen13_13	169 / 3328	125	**122**	125	123	124
queen14_14	196 / 4186	145	141	143	141	142.7
queen15_15	225 / 5180	167	**163**	167	164	166.3
queen16_16	256 / 6320	191	186	205	186	187.7
fpsol2.i.1	269 / 11654	66	66	66	66	66
fpsol2.i.2	363 / 8691	31	31	31	31	31
fpsol2.i.3	363 / 8688	31	31	31	31	31
inithx.i.1	519 / 18707	56	56	56	56	56
inithx.i.2	558 / 13979	35	35	**31**	35	35
inithx.i.3	559 / 13969	35	35	**31**	35	35.3
miles1000	128 / 3216	49	49	49	49	49
miles1500	128 / 5198	77	77	77	77	77
miles250	125 / 387	9	9	9	9	9
miles500	128 / 1170	22	22	22	23	24.3
miles750	128 / 2113	37	36	37	36	37
mulsol.i.1	138 / 3925	50	50	50	50	50
mulsol.i.2	173 / 3885	32	32	32	32	32
mulsol.i.3	174 / 3916	32	32	32	32	32
mulsol.i.4	175 / 3946	32	32	32	32	32
mulsol.i.5	176 / 3973	31	31	31	31	31
myciel3	11 / 20	5	5	5	5	5
myciel4	23 / 71	11	10	10	10	10
myciel5	47 / 236	20	19	19	19	19
myciel6	95 / 755	35	35	35	35	35.7
myciel7	191 / 2360	74	66	**54**	66	67.7
school1	385 / 19095	244	188	-	**184**	203.3
school1_nsh	352 / 14612	192	162	-	**155**	158.7
zeroin.i.1	126 / 4100	50	50	-	50	50
zeroin.i.2	157 / 3541	33	32	-	32	32.3
zeroin.i.3	157 / 3540	33	32	-	32	32.7
le450_5a	450 / 5714	310	256	307	**253**	254.7
le450_5b	450 / 5734	313	254	309	**248**	250
le450_5c	450 / 9803	340	272	315	272	274
le450_5d	450 / 9757	326	278	303	**267**	271.3
le450_15a	450 / 8168	296	272	-	**264**	267.7
le450_15b	450 / 8169	296	**270**	289	271	273.7
le450_15c	450 / 16680	376	359	372	**357**	359.7
le450_15d	450 / 16750	375	360	371	**354**	356
le450_25a	450 / 8260	255	234	255	**221**	227.7
le450_25b	450 / 8263	251	233	251	**228**	229
le450_25c	450 / 17343	355	327	349	327	328.7
le450_25d	450 / 17425	356	336	349	**330**	333.7
dsjc125.1	125 / 736	67	65	64	**60**	60.7
dsjc125.5	125 / 3891	110	109	109	**108**	108.3
dsjc125.9	125 / 6961	119	119	119	119	119
dsjc250.1	250 / 3218	179	173	176	**169**	170.3
dsjc250.5	250 / 15668	233	232	231	**230**	230.3
dsjc250.9	250 / 27897	243	243	243	243	243

Table 2. Algorithms comparison regarding time needed for generation of tree decompositions

Instance	$|V| / |E|$	KBH	TabuS	BB	IHA-best(AVG)	IHA-total(AVG)
anna	138 / 986	1.24	2776.93	1.64	0.1	11.0
david	87 / 812	0.56	796.81	77.6538	0.1	11.0
huck	74 / 602	0.24	488.76	0.041	0.1	11.0
homer	561 / 3258	556.82	157716.56	10800	105.7	206.7
jean	80 / 508	0.29	513.76	0.05	0.1	11.0
games120	120 / 638	5.2	2372.71	-	123.3	224.3
queen5_5	25 / 160	0.04	100.36	5.409	0.1	11.0
queen6_6	36 / 290	0.16	225.55	81.32	0.1	11.0
queen7_7	49 / 476	0.51	322.4	543.3	0.1	11.0
queen8_8	64 / 728	1.49	617.57	10800	17.7	118.7
queen9_9	81 / 1056	3.91	1527.13	10800	1.0	102.0
queen10_10	100 / 1470	9.97	3532.78	10800	5.3	106.3
queen11_11	121 / 1980	23.36	5395.74	10800	11.0	112.0
queen12_12	144 / 2596	49.93	10345.14	10800	18.3	119.3
queen13_13	169 / 3328	107.62	16769.58	10800	30.7	131.7
queen14_14	196 / 4186	215.36	29479.91	10800	834.9	3835.0
queen15_15	225 / 5180	416.25	47856.25	10800	249.3	3250.0
queen16_16	256 / 6320	773.09	73373.12	10800	182.2	3183.0
fpsol2.i.1	269 / 11654	319.34	63050.58	0.587076	6.7	17.7
fpsol2.i.2	363 / 8691	8068.88	78770.05	0.510367	11.0	22.0
fpsol2.i.3	363 / 8688	8131.78	79132.7	0.492061	6.7	17.7
inithx.i.1	519 / 18707	37455.1	101007.52	26.3043	10.7	21.7
inithx.i.2	558 / 13979	37437.2	121353.69	0.05661	12.7	23.7
inithx.i.3	559 / 13969	36566.8	119080.85	0.02734	10.7	21.7
miles1000	128 / 3216	14.39	5696.73	10800	29.3	130.3
miles1500	128 / 5198	29.12	6290.44	6.759	1.0	12.0
miles250	125 / 387	10.62	1898.29	1.788	5.7	16.7
miles500	128 / 1170	4.37	4659.31	1704.62	771.8	3772.0
miles750	128 / 2113	8.13	3585.68	10800	9.7	110.7
mulsol.i.1	138 / 3925	240.24	3226.77	1.407	0.1	11.0
mulsol.i.2	173 / 3885	508.71	12310.37	3.583	0.3	11.3
mulsol.i.3	174 / 3916	527.89	9201.45	3.541	0.7	11.7
mulsol.i.4	175 / 3946	535.72	8040.28	3.622	1.0	12.0
mulsol.i.5	176 / 3973	549.55	13014.81	3.651	1.0	12.0
myciel3	11 / 20	0	72.5	0.059279	0.1	11.0
myciel4	23 / 71	0.02	84.31	0.205	0.1	11.0
myciel5	47 / 236	2	211.73	112.12	0.1	11.0
myciel6	95 / 755	29.83	1992.42	10800	0.3	11.3
myciel7	191 / 2360	634.32	19924.58	10800	11.0	22.0
school1	385 / 19095	41141.1	137966.73	-	2105.4	4794.2
school1_nsh	352 / 14612	2059.52	180300.1	-	3006.3	4885.8
zeroin.i.1	126 / 4100	17.78	2595.92	-	0.1	11.0
zeroin.i.2	157 / 3541	448.74	4825.51	-	42.7	143.7
zeroin.i.3	157 / 3540	437.06	8898.8	-	3.3	104.3
le450_5a	450 / 5714	7836.99	130096.77	10800	2336.3	4789.3
le450_5b	450 / 5734	7909.11	187405.33	10800	3641.7	5001.0
le450_5c	450 / 9803	103637.17	182102.37	10800	1057.3	3947.0
le450_5d	450 / 9757	96227.4	182275.69	10800	735.3	3736.3
le450_15a	450 / 8168	6887.15	117042.59	-	3235.0	4942.0
le450_15b	450 / 8169	6886.84	197527.14	10800	4073.0	5001.0
le450_15c	450 / 16680	122069	143451.73	10800	2446.3	4599.7
le450_15d	450 / 16750	127602	117990.3	10800	3359.3	5001.0
le450_25a	450 / 8260	4478.3	143963.41	10800	2629.7	4739.3
le450_25b	450 / 8263	4869.97	184165.21	10800	3039.3	4555.3
le450_25c	450 / 17343	10998.68	151719.58	10800	3737.3	5001.0
le450_25d	450 / 17425	11376.02	189175.4	10800	2911.0	5001.0
dsjc125.1	125 / 736	171.54	1532.93	10800	696.7	3697.7
dsjc125.5	125 / 3891	38.07	2509.97	10800	1.3	12.3
dsjc125.9	125 / 6961	55.6	1623.44	260.879	0.1	11.0
dsjc250.1	250 / 3218	5507.86	28606.12	10800	1554.3	4115.7
dsjc250.5	250 / 15668	1111.66	14743.35	10800	351.7	3352.7
dsjc250.9	250 / 27897	1414.58	30167.7	10800	0.3	11.3

(IHA-best(AVG)) and the time of the overall run of algorithm (IHA-total (AVG)) in each instance (AVG indicates that the average over three runs is taken). IHA algorithm stops for easy instances after 10 seconds of non improvement of solution, for middle instances after 100 seconds, and for harder instances after 3000 seconds of non improvement of solution. The maximal running time of algorithm for each instance is set to be 5000 seconds.

Based on the results given in Tables 1 and 2 we conclude that our algorithm gives better results for 35 instances compared to [9] for the upper bound of treewidth, whereas algorithm in [9] gives better results than our algorithm for 1 problem. Compared to the algorithm proposed in [3] our approach gives better upper bounds for 17 instances, whereas algorithm in [3] gives better upper bounds for 5 instances. Further, compared to branch and bound algorithm proposed in [6] our algorithm gives better upper bounds for treewidth for 24 instances, whereas the branch and bound algorithm gives better results compared to our algorithm for 4 instances. Considering the time, a direct comparison of algorithms can not be done, as the algorithms are executed in computers with different processors and memory. However, as we can see based on the results in Table 2 our algorithm gives good time performance and for some instances it decreases significantly the time needed for generation of tree decompositions. Based on our experiments the efficiency of our algorithm is due to applying of LS1 algorithm in the construction phase of IHA. In LS1 only one solution is evaluated during each iteration. When using LS2 the number of solutions to be evaluated during most of iterations is much larger.

4   Conclusions

In this paper, we proposed a new heuristic algorithm for finding an upper bound of tree decompositions for a given undirected graph. The proposed algorithm has been applied in different DIMACS vertex coloring instances. The results show that our algorithm achieves very good results for the upper bound of treewidth for different instances. In particular the algorithm improves the best existing treewidth upper bounds for 17 instances, and it has a good time performance.

For the future work we are considering further improvement of proposed algorithm by automatic adaptation of different parameters such as the acceptance criteria, perturbation mechanism, and other parameters in the local search procedure. Additionally we plan to extend the existing algorithm for generation of hypertree decompositions.

Acknowledgments. This paper was supported by the Austrian Science Fund (FWF) project: *Nr. P17222-N04, Complementary Approaches to Constraint Satisfaction.*

References

1. S. Arnborg, D. G. Corneil, and A. Proskurowski. Complexity of finding embeddings in a k-tree. *SIAM J. Alg. Disc. Meth.*, 8:277–284, 1987.
2. H. L. Bodlaender. Discovering treewidth. *technical report UU-CS-2005-018, Utrecht University*, 2005.
3. F. Clautiaux, A. Moukrim, S. Négre, and J. Carlier. Heuristic and meta-heurisistic methods for computing graph treewidth. *RAIRO Oper. Res.*, 38:13–26, 2004.
4. D. R. Fulkerson and O.A. Gross. Incidence matrices and interval graphs. *Pacific Journal of Mathematics*, 15:835–855, 1965.
5. F. Gavril. Algorithms for minimum coloring, maximum clique, minimum coloring cliques and maximum independent set of a chordal graph. *SIAM J. Comput.*, 1:180–187, 1972.
6. Vibhav Gogate and Rina Dechter. A complete anytime algorithm for treewidth. *In Proceedings of the 20th Annual Conference on Uncertainty in Artificial Intelligence, UAI-04*, pages 201–208, 2004.
7. D. S. Johnson and M. A. Trick. The second dimacs implementation challenge: Np-hard problems: Maximum clique, graph coloring, and satisfiability. *Series in Discrete Mathematics and Theoretical Computer Science, American Mathematical Society*, 1993.
8. U. Kjaerulff. Optimal decomposition of probabilistic networks by simulated annealing. *Statistics and Computing*, 1:2–17, 1992.
9. A. Koster, H. Bodlaender, and S. van Hoesel. Treewidth: Computational experiments. *Electronic Notes in Discrete Mathematics 8, Elsevier Science Publishers*, 2001.
10. P. Larranaga, C.M.H Kujipers, M. Poza, and R.H. Murga. Decomposing bayesian networks: triangulation of the moral graph with genetic algorithms. *Statistics and Computing (UK)*, 7(1):1997, 1991.
11. N. Robertson and P. D. Seymour. Graph minors. ii. algorithmic aspects of treewidth. *Journal Algorithms*, 7:309–322, 1986.
12. K. Shoikhet and D. Geiger. A practical algorithm for finding optimal triangulations. *In Proc. of National Conference on Artificial Intelligence (AAAI'97*, pages 185–190, 1997.
13. R.E. Tarjan and M. Yannakakis. Simple linear-time algorithm to test chordality of graphs, test acyclicity of hypergraphs, and selectively reduce acyclic hypergraphs. *SIAM J. Comput.*, 13:566–579, 1984.

Edge Assembly Crossover for the Capacitated Vehicle Routing Problem

Yuichi Nagata

Graduate School of Information Sciences,
Japan Advanced Institute of Science and Technology
nagatay@jaist.ac.jp

Abstract. We propose an evolutionary algorithm (EA) that applies to the capacitated vehicle routing problem (CVRP). The EA uses edge assembly crossover (EAX) which was originally designed for the traveling salesman problem (TSP). EAX can be straightforwardly extended to the CVRP if the constraint of the vehicle capacity is not considered. To address the constraint violation, the penalty function method with *2-opt* and *Interchange* neighborhoods is incorporated into the EA. Moreover, a local search is also incorporated into the EA. The experimental results demonstrate that the proposed EA can effectively find the best-known solutions on *Christofides* benchmark. Moreover, our EA found ten new best solutions for *Golden* instances in a reasonable computation time.

1 Introduction

The Vehicle Routing Problem (VRP) is a practical problem that has been widely studied in the OR community. Let $G = (V, E)$ be a complete undirected graph with a set of $n + 1$ vertices $V = \{v_0, v_1, \ldots, v_n\}$ and a set of edges E. v_0 is a *depot* and v_i ($i \in \{1, \ldots, n\}$) represent n customers, each having non-negative demand q_i. Each edge (v_i, v_j) has non-negative distance c_{ij}. The VRP is to find a set of m routes of minimum total distance, such that each route starts and ends at the depot, each customer is visited exactly once, the total demand of any route does not exceed Q (vehicle capacity constraint). Note that m is also a decision variable. The definition described here is the most basic type of the VRP, that is called the capacitated VRP (CVRP). There are several variants of the VRP with additional constraints. In this paper, we focus on the CVRP.

Due to its theoretical and practical interest, the VRP has received a great amount of attentions since its proposal in the 1950's. In recent years, great success in finding near-optimal solutions has been achieved via meta-heuristics including deterministic annealing [1], tabu search [2][3][4], evolution strategy [5] and population based search [6][7], etc. These results are reviewed in [8].

The traveling salesman problem (TSP) is known to be a special case of the CVRP where the vehicle capacity Q is infinity. In despite of the similarity of the definitions between the TSP and the CVRP, difficulties of these problems are quite different. For example, exact algorithms can only solve relatively small

C. Cotta and J. van Hemert (Eds.): EvoCOP 2007, LNCS 4446, pp. 142–153, 2007.

CVRP instances ($n < 100$) while some of non-trivial relatively large TSP instances ($n < 30,000$) can be solved [9]. Before now, many heuristic methods have been applied to the TSP and these techniques are sometimes useful for solving the VRP. One powerful operator (crossover) developed in population based searches for the TSP is edge assembly crossover (EAX) [10]. In recent study [11][12], evolutionary algorithms (EAs) using EAX could find several optimal solutions up to 24,000 nodes within a day.

In this paper, an evolutionary algorithm using EAX is applied to the CVRP. And we demonstrate that EAX has a great potential for solving the CVRP. EAX can be straightforwardly extended to the CVRP. However, EAX generates children without respect to the vehicle capacity constraints, that results in producing infeasible solutions. Such solutions are modified by local improvements based on a penalty function that penalize the overhead of the vehicle capacity. Well-known *2-opt* and *Interchange* neighborhoods [13] are used in a modification procedure. Moreover, a local search is also incorporated into the EA because local searchs are almost mandatory to achieve high quality solutions for the VRP [7].

The remainder of this paper is organized as follows. In Section 2, we briefly introduce EAX designed for the TSP. And then we propose EAX designed for the CVRP in Section 3. An EA including EAX as a crossover operator and incorporated techniques such as a modification procedure and a local search are described in Section 4. In Section 5, computational results are presented. Finally, we present our conclusions in Section 6.

2   EAX for the TSP

In this section, we briefly look at EAX crossover designed for the TSP [10] without introducing the EA. Fig. 1 illustrates a rough outline of EAX designed for the TSP. EAX generates new tours (children) by combining two tours (parents) in a population. EAX consists of two phases.

In the first phase, intermediate solutions are constructed by assembling edges from parents under a relaxed condition of the TSP. In this relaxation, the constraint imposed on intermediate solutions is that $\delta^C(v) = 2$ ($v \in V$) where $\delta^C(v)$ denotes the number of edges incident to a vertex v in an intermediate solution C. Therefore, each intermediate solution generally consists of several sub-tours. In the second phase, each intermediate solution is modified into a feasible solution (tour) by merging sub-tours. In this modification, any two sub-tours are merged by deleting one edge from each of the two sub-tour and adding the edges to connect them. Which sub-tours are connected and which edges are deleted are determined so as to minimize a resulting tour length.

EAX has two advantages; (i) a wide variety of children can be generated from a pair of parents because intermediate solutions are constructed under the relaxed condition of the TSP, and (ii) children can be constructed without introducing long edges. So, we can intuitively expect that EAX work well for solving the TSP.

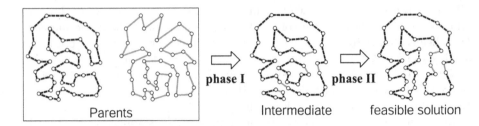

Fig. 1. EAX crossover for the TSP

3 EAX for the CVRP

In this section, we propose EAX designed for the CVRP. A basic algorithm is almost the same as that for the TSP [11][12].

3.1 Algorithm of EAX

EAX designed for the TSP can be straightforwardly extended to the CVRP when the vehicle capacity constraint is neglected. Algorithm 1 and Fig. 2 illustrate an algorithm of EAX designed for the CVRP. EAX finally generates intermediate solutions (II) from parents (indi-A and indi-B). For the sake of simplicity, let indi-A and indi-B have the same number of routes, that is denoted by m. And let all customer demands q_i be 1, meaning that each feasible route cannot includes more than Q customers. In the example illustrated in Fig. 2, $m = 3$ and $Q = 15$.

EAX designed for the CVRP also consist of two phases as is the case of the TSP. In the first phase (step (1) – step (4)), intermediate solutions (I) are constructed by assembling edges from parents under a relaxed condition of the CVRP. In this relaxation, the constraint imposed on intermediate solutions (I) is that $\delta^C(v_0) = 2m$ and $\delta^C(v) = 2$ ($v \in V \backslash \{v_0\}$). Therefore, each intermediate solution (I) consists of m routes (cycles including the depot), and sub-tours (cycles not including the depot). In the second phase (step 5), sub-tours are merged into routes without introducing long edges.

However, resulting intermediate solutions (II) generally violate the vehicle capacity constraint. In this example, (a") and (b") are infeasible solutions, each including infeasible routes of the total demands of 22 and 17, respectively. Therefore, such infeasible solutions must be modified into feasible solutions. A modification procedure is not included in the procedure of EAX and this procedure is described in 4.2.

Before describing an algorithm of EAX, some notations are defined. E_A and E_B are defined as sets of edges included in indi-A and indi-B, respectively. $G_{AB} = (V, E_A \cup E_B \backslash (E_A \cap E_B))$. *AB-cycle* is defined as a cycle on G_{AB} such that indi-A's edges and indi-B's edges are linked alternately. *E-set* is defined as any combination of *AB-cycles*. Although G_{AB} can be defined as $(V, E_A \cup E_B)$, the former definition is employed in this paper.

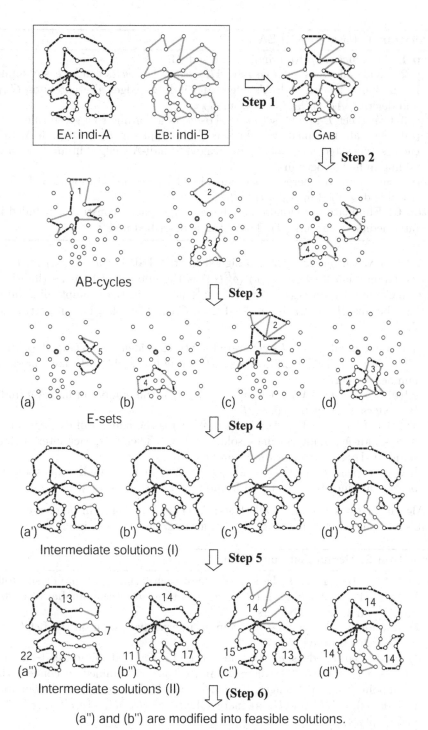

Fig. 2. Basic steps of EAX for the VRP

Algorithm 1. Basic steps of EAX

Step 1 Construct G_{AB} from indi-A and indi-B.

Step 2 Divide all edges on G_{AB} into *AB-cycles*. *AB-cycles* can be easily found by randomly tracing indi-A's edges and indi-B's edges alternately on G_{AB} and deleting *AB-cycles* found from G_{AB}.

Step 3 Construct *E-sets* by selecting *AB-cycles* according to a given rule.

Step 4 Generate intermediate solutions (I) by applying *E-sets* to indi-A, *i.e.*, each is generated form indi-A by removing indi-A's edges in an *E-set* and adding indi-B's edges in an *E-set*.

Step 5 Sub-tours are merged into routes in each intermediate solution (I). The detail is described in Algorithm 2.

(Step 6) Eliminate the violation of the vehicle capacity constraint included in intermediate solutions (II). The detail is described in 4.2.

A minor variation from EAX designed for the TSP is that $\delta_G^A(v_0)$ $(\delta_G^B(v_0))$ can be larger than 2 where $\delta_G^A(v)$ $(\delta_G^B(v))$ is the number of indi-A's (indi-B's) edges incident to v on G_{AB}. Note that $\delta_G^A(v_0) = 1$ in the example illustrated in Fig 2 because $E_A \cap E_B$ is removed from G_{AB}. The following properties are satisfied.

- The union of all *AB-cycles* generated in Step 2 is identical to G_{AB} if $\delta_G^A(v) = \delta_G^B(v)$ $(v \in V)$. This condition is satisfied if indi-A and indi-B have the same number of routes.
- $\delta_E^A(v) = \delta_E^B(v)$ $(v \in V)$ where $\delta_E^A(v)$ $(\delta_E^B(v))$ is the number of indi-A's (indi-B's) edges incident to v in an *E-set*.
- $\delta^C(v) = \delta^A(v)$ $(v \in V)$ where $\delta^A(v)$ $(\delta^C(v))$ is the number of edges incident to v in indi-A (an intermediate solution (I) C). Therefore, each intermediate solution (I) consists of m routes and sub-tours.
- If G_{AB} is defined as $(V, E_A \cup E_B \backslash (E_A \cap E_B))$, common edges between the parents are inevitably included in intermediate solutions (I).

Algorithm 2 describes a detail of Step 5 where sub-tours can be merged into routes without introducing long edges in an intermediate solution (I).

Algorithm 2. Merging sub-tour (Step 5)

(5-1) Let U_i $(i = 1, \ldots, k)$ be sets of edges, each forming a route or sub-tour in an intermediate solution (I) where k is the sum of the number of routes and the number of sub-tours.

(5-2) Randomly select a sub-tour from an intermediate solution (I). Let U_r be a selected sub-tour.

(5-3) Find a pair of edges, $e \in U_r$ and $e' \in U_j$ $(j \neq r)$ that minimizes $\{-w(e) - w(e') + w(e'') + w(e''')\}$ where e'' and e''' are determined to connect the breakpoints. $w(e)$ is a length of a edge e. Let U_s be a route or sub-tour that includes edge e'. U_r and U_s are merged by $U_r := (U_r \cup U_s - \{e, e'\}) \cup \{e'', e'''\}$, and empty U_s.

(5-4) If there is no sub-tour, then terminate, else go to (5-2).

3.2   Methods of Constructing E-Sets

In Step 3, *E-sets* can be constructed of any combination of *AB-cycles* and various intermediate solutions can be generated depending on *E-sets*. One simple method for constructing *E-sets* is selecting *AB-cycles* randomly, that is called EAX-Rand. Although EAX-Rand can generate a wide variety of children, the following two methods for selecting *AB-cycles* are known to be better than EAX-Rand in the studies of the TSP [11][12].

EAX-1AB: An *E-set* is constructed of a single *AB-cycle*. For example, *E-sets* (a) and (b) can be constructed (See Fig. 2).

EAX-Block: First, randomly select a single *AB-cycle*. Let it be a center *AB-cycle*. In addition, select *AB-cycles* incident to the center *AB-cycle*. For example, *E-set* (c) can be constructed where *AB-cycle* 2 is selected as a center *AB-cycle* (See Fig. 2).

Intermediate solutions (I) generated by EAX-1AB tends to be similar to indi-A because these are formed from indi-A by removing a relatively small number of indi-A's edges and adding the same number of indi-B's edges. On the other hand, EAX-Block can generate intermediate solutions (I) by assembling a block (a set of relatively large number of geographically close edges) of indi-A's edges and a block of indi-B's edges. EAX-1AB should be used in an initial stage of an EA because relatively localized moves can effectively improve solution candidates in this stage. EAX-Block should be used after EAX-1AB can no longer improve solution candidates because these solutions get stuck in deep local optima.

4   EA for Solving the CVRP

In this section, an EA for solving the CVRP is presented. First, we describe a skeleton of the proposed EA that includes EAX as a crossover operator. Next, sub-functions used in the proposed EA are described.

4.1   Main Procedure

Procedure EA is a pseudo-code of the proposed EA. An initial population is generated by Creat_Initial_Solution() (line 2). For each pair of parents (denoted by p_A and p_B) (line 7), N_{ch} children are generated (line 9 – 14) and the child having the shortest total distance replaces the individual in the population selected as p_A if it is better than p_A (line 8,13,15). This replacement is suitable to maintain the population diversity because children generated by EAX-1AB tends to be similar to p_A. If children generated by EAX violate the vehicle capacity constraint, the violation is eliminated by a modification procedure (line 11). Moreover, feasible solutions are locally optimized by a local search (line 12). When using EAX (line 10), two types of EAX described in 3.2 are used. Initially, EAX-1AB is used until the best solution in the population stagnates over 20 generations. After that, EAX-Block is used until the best solution in the population stagnates over 50 generations. Then, terminate a run (line 17).

Procedure EA(N_{pop}, N_{ch})
begin
1 :**for** $i := 1$ **to** N_{pop} **do**
2 : $indi_i :=$ Creat_Initial_Solution();
3 :**end for**
4 :**repeat**
5 :   indexes $i \in \{1, \ldots, N_{pop}\}$ are randomly assigned to the population;
6 :   **for** $i := 1$ **to** N_{pop} **do**
7 : $p_A := indi_i$; $p_B := indi_{i+1}$;
8 : $c_{best} := p_A$;
9 :   **for** $j := 1$ **to** N_{ch} **do**
10:            $c :=$ EAX(p_A, p_B); // EAX-1AB or EAX-Block
11:            $c :=$ Modification(c);
12:            $c :=$ Local_Search(c);
13:            **if** $c.distance < c_{best}.distance$ **then** $c_{best} := c$;
14:        **end for**
15:        $indi_i := c_{best}$;
16:    **end for**
17:**until** termination condition is satisfied;
18:**return** best individual in the population;
end

4.2   Sub-functions

Local Search. We use the following two neighborhoods that are known to be suitable for the VRP [13]. *2-Opt* neighborhood is defined as a set of all possible solution candidates that are obtained from a current solution by deleting two edges and adding new two edges to connect the breakpoints in other possible ways. *Interchange* neighborhood is defined as a set of all possible solution candidates obtained from a current solution by interchanging two segments between different routes in other possible way. Fig. 3 illustrates several cases.

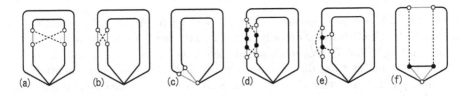

Fig. 3. Neighborhoods for the VRP. Points represent successive customers on routes. Gray and dotted edges mean removed and added edges, respectively. (a), (b) and (c) are examples of *2-opt*. (d), (e) and (f) are examples of *Interchange*.

Functions 2-Opt(c) and Interchange(c, λ, μ) are defined as the local searches that return local optima using *2-Opt* and *Interchange* (λ and μ mean the number of vertices of the swapped segments) neighborhoods, respectively. These local searches are executed with the first improvement. Procedure Local_Search(c) is a pseudo-code of the local search used in the proposed EA.

Procedure Local_Search(c)
begin
1: $c :=$ 2-Opt(c);
2: $c :=$ Interchange($c, 0, 1$);
3: $c :=$ Interchange($c, 1, 1$);
4: **return** c;
end

Modification. Procedure Modification(c) modifies infeasible solution candidates that violate the vehicle capacity constraint. In the modification process, the penalty function method is used. In the studies of the CVRP, the following penalty function is simple and is frequently used [7] where $F_{VRP}(c)$ is the total distance of routes, and overcap(c) is the overhead of the capacity (the sum of the excess of demand on routes) in a solution candidate c.

$$F_{penalty}(c) = F_{\mathrm{VRP}}(c) + \alpha \cdot \mathrm{overcap}(c)$$

In the procedure, a route r violating the constraint is randomly selected (line 2). And then, the best solution candidate (evaluated by the penalty function) in restricted *2-Opt* and *Interchange* neighborhoods are searched (line 3– 5). For example, function 2-Opt_BestImp(c, r) returns the best solution candidate in the restricted *2-Opt* neighborhood of a solution candidate c where one deleting edge must be included in the route r. Function Interchange_BestImp($c, \lambda, 0, r$) is also defined in a similar fashion where a segment of length λ must be included in the route r. These processes are repeated until a feasible solution is obtained.

In addition, neighboring solutions that does not decrease overcap(c) are neglected in the modification process so that feasible solutions can be obtained robustly. The parameter α should be determined appropriately. A method of determining α is described next.

Procedure Modification(c)
// Penalty function is used in this process.
Let Q_r ($r = 1, \ldots, m$) be the sum of demand on a route r.
begin
1: **repeat**
2:     Randomly select $r \in \{1, \ldots, m\}$ such that $Q_r > Q$;
3:     $c_1 :=$ 2-Opt_BestImp(c, r);
4:     $c_2 :=$ Interchange_BestImp($c, r, 1, 0$);
5:     $c_3 :=$ Interchange_BestImp($c, r, 2, 0$);
6:     $c :=$ Select_Best_Evaluated(c_1, c_2, c_3);
7: **until** $Q_r \leq Q$ ($r = 1, \ldots, m$);
8: **return** c;
end

Preprocessing. Procedure Creat_Initial_Solution() generates feasible solutions. In this process, function Initial_Solution() (line 1) produces identical feasible solutions, each defined by $\{v_0, v_i, v_0\}$ ($i \in V \setminus \{v_0\}$). During the local search using the penalty function (line 2), routes in a temporal solution can be merged due to the special cases of *2-Opt* and Interchange neighborhoods. The number of routes in each solution candidate generated by this process (line 2) tend to be small with decreasing the parameter α (If α is 0, the optimum is a tour). If solution candidates generated by this process (line 2) are infeasible, such solutions are modified by the modification process (line 3) described above. Moreover, feasible solutions are locally optimized (line 4).

The parameter α is determined so that Procedure Creat_Initial_Solution() can steady generate relatively better solution candidates. Let m be the number of routes of the relatively better solutions. We employed only solution candidates consisting of m routes as the initial population.

Procedure Creat_Initial_Solution()
begin
1: $c :=$ Initial_Solution();
2: $c :=$ Local_Search(c); // Penalty function is used.
3: $c :=$ Modification(c);
4: $c :=$ Local_Search(c); // Normal evaluation is used.
5: **return** c;
end

5   Experimental Results

In this section, we describe experimental results of the proposed EA. And then, the results are compared with other heuristic methods that are known to be current best heuristics on the CVRP.

The proposed EA is applied to well-known CVRP benchmarks. *Christofides* benchmark [14] consisting of 14 instances ($50 \leq n \leq 199$) was proposed in 1979. *Golden* benchmark [1] consisting of relatively large 20 instances ($200 \leq n \leq 480$) was proposed in 1998. In Table 1, *Christofides* and *Golden* instances are denoted by C# and G#, respectively. Customers in C1–C5 are randomly distributed, whereas customers in C11 and 12 are clustered together. Customers in G9–G20 are arranged in distinct geometric shapes. We excluded C6–C10 and G1–G8 from our experiments because an additional constraint called *total duration constraint* is imposed on these instances and there is not enough room to address these instances.

The proposed EA is implemented in C++ and executed on Xeon 3.2 GHz single processor with a parameter setting: $N_{pop} = 100$, $N_{ch} = 30$. Ten runs were executed for each instance.

Table 1 lists the results of the proposed EA. The column headings are as follows: "Best" and "Average" mean the best and the averaged solution values, respectively, "#Better" means the number of trials that found best-known

Table 1. Computational results of the proposed EA (Ten runs)

Instances	n	α	Best	#Better	Average	Err.	Seconds[1]	Pre. Best
C1	50	0.3	**524.61**	10	524.61	0.00	12	524.61
C2	75	0.5	**835.26**	10	835.26	0.00	63	835.26
C3	199	0.5	**826.14**	10	826.14	0.00	31	826.14
C4	150	0.5	**1028.42**	9	1028.51	0.01	136	1028.42
C5	199	0.3	**1291.29**	2	1293.93	0.20	720	1291.29
C11	120	0.8	**1042.11**	10	1042.11	0.00	30	1042.11
C12	100	0.5	**819.56**	10	819.56	0.00	5	819.56
G9	255	0.05	**580.60***	10	582.34	-0.18	1552	583.39
G10	323	0.05	**738.92***	8	740.91	-0.15	2508	742.03
G11	399	0.1	**917.17***	6	918.19	-0.03	3835	918.45
G12	483	0.07	1108.48	0	1110.71	0.32	6801	1107.19
G13	252	0.04	**857.19***	7	858.84	-0.03	1165	859.11
G14	320	0.05	**1080.55***	8	1080.93	-0.04	1620	1081.31
G15	396	0.05	**1340.24***	8	1344.02	-0.09	1924	1345.23
G16	480	0.3	2171.30	0	2178.78	34.27	3219	1622.69
G17	240	0.2	**707.76***	10	707.77	-0.00	718	707.79
G18	300	0.2	**995.39***	10	996.62	-0.21	1261	998.73
G19	360	0.3	**1366.14***	3	1367.31	0.03	2013	1366.86
G20	420	0.3	**1820.54***	1	1822.97	0.10	3169	1821.15

(1) Xeon 3.2 GHz

or new best solutions, "Err." means the averaged percentage excess from the best-known solution values, "Seconds" means the averaged CPU time in seconds required for a single run, "Pre Best" means the best-known solution values [8][15]. In the column "Best", solution values are in boldface if these values are equal or better than the best-known values, and "*" means new best solutions.

Table 2 lists the results of recent CVRP heuristics reported in the review paper for the CVRP [8]. Three heuristics demonstrating relatively better results are listed. AGES [5] is known to be the current best meta-heuristic for the CVRP (and the VRP with time window). Prins's method [6] is frequently referred as a good memetic algorithm for the CVRP. GTS [2] is a kind of tabu search algorithm. These results correspond to a single run unlike in the case of our experiments. In this table, "Value" means the solution value obtained by a single run, "Minutes" means the CPU time in minutes required for a single run (Note that "Seconds" is used in Table 1). "Pre Best" means the best-known solution values where these values for the *Golden* instances were obtained by AGES itself.

As shown in Table 1, the proposed EA can find the best-known solutions in all *Christofides* instances employed in our experiments. Indeed, the proposed EA can easily find the best-known solutions except for C5 instance even if the parameters N_{pop} and N_{ch} are set to be 50 and 10, respectively. On the other hand, C5 is a hard problem because other heuristics except for AGES could not find the best-known solution and AGES need a long time to find it. Although the proposed EA found the best-known solution only two times over ten trials, the CPU time for a single run is smaller than that of AGES.

Table 2. Computational results of other heuristics for the VRP (A single run)

Instances	n	GTS Toth and Vigo (2003)			Prins(2004)			AGES best Mester and Braysy (2004)			Pre. Best
		Value	Err.	Minutes[2]	Value	Err.	Minutes[3]	Value	Err.	Minutes[4]	
C1	50	**524.61**	0.00	0.81	**524.61**	0.00	0.01	**524.61**	0.00	0.01	524.61
C2	75	838.60	0.40	2.21	**835.26**	0.00	0.77	**835.26**	0.00	0.26	835.26
C3	199	828.56	0.29	2.39	**826.14**	0.00	0.46	**826.14**	0.00	0.05	826.14
C4	150	1033.21	0.47	4.51	1031.63	0.31	5.50	**1028.42**	0.00	0.47	1028.42
C5	199	1318.25	2.09	7.50	1300.23	0.69	19.10	**1291.29**	0.00	101.93	1291.29
C11	120	1042.87	0.07	3.18	**1042.11**	0.00	0.30	**1042.11**	0.00	0.05	1042.11
C12	100	**819.56**	0.00	1.10	**819.56**	0.00	0.05	**819.56**	0.00	0.63	819.56
G9	255	593.35	1.71	11.67	591.54	1.40	14.32	**583.39**	0.00	8.33	583.39
G10	323	751.66	1.30	15.83	751.41	1.26	36.58	**742.03**	0.00	6.00	742.03
G11	399	936.04	1.92	33.12	933.04	1.59	78.50	**918.45**	0.00	110.00	918.45
G12	483	1147.14	3.61	42.90	1133.79	2.40	30.87	**1107.19**	0.00	600.00	1107.19
G13	252	868.80	1.13	11.43	875.16	1.87	15.30	**859.11**	0.00	10.25	859.11
G14	320	1096.18	1.38	14.51	1086.24	0.46	34.07	**1081.31**	0.00	1.22	1081.31
G15	396	1369.44	1.80	18.45	1367.37	1.65	110.48	**1345.23**	0.00	7.17	1345.23
G16	480	1652.32	1.83	23.07	1650.94	1.74	130.97	**1622.69**	0.00	20.00	1622.69
G17	240	711.07	0.46	14.29	710.42	0.37	5.86	**707.79**	0.00	0.75	707.79
G18	300	1016.83	1.81	21.45	1014.80	1.61	39.33	**998.73**	0.00	2.50	998.73
G19	360	1400.96	2.49	30.06	1376.49	0.70	74.25	**1366.86**	0.00	6.00	1366.86
G20	420	1915.83	5.20	43.05	1846.55	1.39	210.42	**1821.15**	0.00	8.40	1821.15

(2) Pentium 200MHz, (3) Pentium III 1GHz, (4) Pentium IV 2GHz

Surprisingly, the proposed EA found new best solutions for ten *Golden* instances except for G12 and G16 [1]. Moreover, the numbers of trials that found the best-known or new best solutions are larger than six in eight instances. Although the CPU times of AGES tend to be smaller than those of the proposed EA, no date is available when more CPU times are allowed in AGES. The number of routes of the new best solutions are equal to those of the previous best-known solutions, respectively. Various new best solutions were obtained in each instances.

Additionally the proposed EA found a new best solution of total distance of 24396.38 for well-known tai385 (n = 385) instance included in *Taillard* benchmark [15]. The previous best-known solution of length 24431.44 had not been improved since 1995.

6   Conclusion and Future Work

In this paper, we proposed the EA that is applied to the CVRP. The main idea is applying EAX crossover to the CVRP. EAX can be straightforwardly extended to the CVRP if the vehicle capacity constraint is not considered. Infeasible solutions generated by EAX are modified by the modification procedure consisting of the penalty function method with *2-opt* and *Interchange* neighborhoods.

---

[1] The solution values for instance G16 is significantly worse. In my opinion, the highly clustered customers arrangement in this instance is related to this result.

We demonstrated that the proposed EA could find ten new best solutions on well-known *Golden et al.* benchmark for the CVRP (Table 1). In my opinion, EAX has the following features in solving the CVRP; (i) intermediate solutions (II) (infeasible solutions) can be constructed of short edges, and (ii) intermediate solutions (II) can be modified into feasible solutions by relatively local moves even though the overheads of the capacity are large (Note that such infeasible solutions may be neglected when using usual penalty function methods).

We believe that EAX has a great potential for developments of the studies of the VRP including more complex constraints such as the VRP with time window (VRPTW). However, the proposed EAX can not work well on the VRPTW because EAX does not consider the order of customers in routes. We are now extended the proposed EAX to the VRPTW.

References

1. B.L. Golden, E.A. Wasil, J.P. Kelly, I.M. Chao. Metaheuristics in vehicle routing. In T.G Crainic and C. Laporte, editors, Fleet Management and Logistics, Kiuwer, Boston, MA, pp. 33–56, 1998.
2. P. Toth and D. Vigo. The granular tabu search and its application to the Vehicle Routing problem, INFORMS Journal on Computating 15, pp. 333– 346.
3. E. D. Taillard. Parallel Iterative Search Methods for Vehicle Routing Problems, Networks 23, 661–673, 1993.
4. J. Kelly, J. P. Xu. A Network Flow-Based Tabu Search Heuristic for the Vehicle Routing Problem, Transportation Science 30, 379–393, 1996.
5. D. Mester and O. Braysy. Active Guided Evolution Strategies for Large Scale Vehicle Routing Problems with Time Windows, Computers & Operations Research 32, pp. 1593–1614, 2005.
6. Prins, C. A simple and effective evolutionary algorithm for the vehicle routing problem, Computers & Operations Research 31, pp. 1985–2002, 2004.
7. E. Alba, B. Dorronsoro. Solving the Vehicle Routing Problem by Using Cellular Genetic Algorithms, Evolutionary Computation in Combinatorial Optimization - EvoCOP 2004, LNCS, Vol. 3004, pp. 11–20, 2004.
8. J. F. Cordeau, M. Gendreau, A. Hertz, G. Laporte, J.S. Sormany. New Heuristics for the Vehicle Routing Problem, Logistics Systems: Design and Optimization, A. Langevin and D. Riopel, editors, Springer, New York, pp. 279–297, 2005.
9. http://www.tsp.gatech.edu/sweden/index.html
10. Y. Nagata, S. Kobayashi. Edge Assembly Crossover: A High-power Genetic Algorithm for the Traveling Salesman Problem, Proc. of the 7th Int. Conference on Genetic Algorithms, pp. 450–457, 1997.
11. Y. Nagata, Fast EAX algorithm Considering Population Diversity for Traveling Salesman Problems, Proc. of the 6th Int. Conf. on EvoCOP2006, pp. 171–182, 2006.
12. Y. Nagata. New EAX crossover for large TSP instances, Proceedings of the 9th International Conference on Parallel Problem Solving from Nature, LNCS, Vol. 4193, pp. 372–381, 2006.
13. A.P. Kindervater, W.P. Savelsbergh, Chapter 10 in Local Search in Combinatorial optimization, edited by E. Aarts and J.K. Lenstra, John Wiley & Son, 1997.
14. N. Christofides, A. Mingozzi, P. Toth. The vehicle routing problem. Chap. 11 in Combinatorial Optimization, edited by the same authors and C. Sandi, Wiley, 1979.
15. http://neo.lcc.uma.es/radi-aeb/WebVRP/

Tackling the Container Loading Problem: A Hybrid Approach Based on Integer Linear Programming and Genetic Algorithms

Napoleão Nepomuceno, Plácido Pinheiro, and André L.V. Coelho

Universidade de Fortaleza, Mestrado em Informática Aplicada
Av. Washington Soares 1321, Sala J-30, Fortaleza, CE, Brazil, 60811-905
napoleao@edu.unifor.br, {placido, acoelho}@unifor.br

Abstract. This paper presents a novel hybrid approach for solving the Container Loading (CL) problem based on the combination of Integer Linear Programming (ILP) and Genetic Algorithms (GAs). More precisely, a GA engine works as a generator of reduced instances for the original CL problem, which are formulated as ILP models. These instances, in turn, are solved by an exact optimization technique (solver), and the performance measures accomplished by the respective models are interpreted as fitness values by the genetic algorithm, thus guiding its evolutionary process. Computational experiments performed on standard benchmark problems, as well as a practical case study developed in a metallurgic factory, are also reported and discussed here in a manner as to testify the potentialities behind the novel approach.

Keywords: Combinatorial Optimization, Hybrid Methods, Container Loading, Integer Linear Programming, Metaheuristics, Genetic Algorithms.

1 Introduction

Roughly speaking, the Container Loading (CL) problem alludes to the task of packing boxes into containers. More precisely, given the dimensions of the containers and the boxes which need to be loaded, the problem can be defined as to find such an arrangement of boxes that optimizes a given objective function – in general, the maximum volume of the loaded boxes. In addition to the geometric constraints, other restrictions can also be considered, such as boxes orientation and cargo stability.

New variants of the CL problem have constantly appeared in the literature, making difficult the development of a single technique to efficiently handle all of them. In [2], two different perspectives on this problem were taken into account. In the first, a particular combination of containers should be chosen in a manner as to completely transport a given cargo. In the second, the larger volume of a given cargo is sought to be allocated into a single container. Anyway, the CL problem can be regarded as a special case of the family of Cutting and Packing (CP) problems [4][5].

In general, the CP problems can be easily formulated and understood, revealing a simplicity that contrasts directly with their real complexity. The importance of this class of problems is well acknowledged in the domains of Computer Science and

C. Cotta and J. van Hemert (Eds.): EvoCOP 2007, LNCS 4446, pp. 154–165, 2007.

Operations Research. In addition, their tackling by recently-proposed methods has provided relevant contributions to the improvement of real-world activities in the industry sector. Due to their inherent difficulty, however, there is still much room for the conception of more sophisticated solution methods to these problems.

A variety of mathematical formulations for CP problems can be easily found in the literature. For instance, in [7], an Integer Linear Programming (ILP) model for a unidimensional cutting stock problem is formulated. Conversely, in [1], a 0-1 ILP model is conceived for a two-dimensional cutting problem. This particular model was later extended for dealing with the 3D case of the CL problem [8]. Even though the theoretical rigor of these models comes to be very appealing, the direct application of exact algorithms for their implementation has proven to be computationally viable only for problem instances of very limited size.

Due to the NP-completeness difficulty inherent to the CP problems, the majority of the successful approaches that have been lately proposed for coping with them have exploited, in a way or another, some source of heuristic information [10]. For instance, the combination of Genetic Algorithms (GAs) with some type of heuristic placement schemes has recently emerged as a promising line of investigation, being addressed by several works in the literature (e.g., see [15] and the references therein).

In this paper, we introduce a novel approach for solving the CL problem, which follows an alternative research direction to those mentioned above. Such direction is more in line with the current trend of hybridizing exact and metaheuristic techniques into unique solutions for Combinatorial Optimization (CO) problems [6]. Overall, the major rationale behind this sort of hybridization lies in the attempt of combining, in an efficient and effective manner, the complementary advantages displayed by both classes of optimization techniques. In practice, in one way, this might provide optimal solutions with shorter execution time; in the other way around, the result would be to obtain novel heuristic solutions more tailored to the specific problem at hand [10].

A general classification of the hybrid methodologies distinguishes them into two main categories: collaborative combinations and integrative combinations [12]. By collaboration, it means that the constituent algorithms exchange information to each other, but none is part of the other. The algorithms may be executed sequentially, intertwined or in parallel. By integration, it means that one technique is a subordinate component of the other. Thus, there is the distinction between a master algorithm – which could be either an exact method or a metaheuristic – and a slave algorithm.

Few applications of this class of hybrid algorithms to CP problems have been investigated yet. One of these works [9] combines interior point algorithms and metaheuristics in order to solve the knapsack problem. By other means, in [11], a combination of genetic and Branch-and-Bound algorithms for the two-dimensional cutting stock problem is described. Moreover, in [13], a different hybrid approach is proposed for the 3D container loading problem, whereby local search techniques are incorporated into an exact algorithm.

In what follows, we present our hybrid approach for tackling the CL problem[1], which is based on a particular framework combining ILP and GA techniques. In fact, it is our feeling that such framework is generic enough for coping well with other

---

[1] More specifically, the CL variant we are considering here can be classified as "3/B/O/" according to the taxonomy proposed by Dyckhoff [4].

types of CP problems and, for this reason, it is introduced in Section 0 through a more broad stance, after which it is specifically realized for the CL case. In the sequel, we present and discuss some simulation results we have achieved by experimenting with a series of benchmarking CL problem instances, as well as with a practical case study developed in a metallurgic factory, in order to assess the effectiveness of the novel approach, contrasting its performance with that achieved by alternative solutions. Some final remarks with indication for future work conclude the paper.

2 Hybrid Framework

The proposed framework prescribes the integration of two distinct conceptual components, according to the setting illustrated in Fig. 1. The first component is named as the Generator of Reduced Instances (GRI), which is in charge of producing reduced problem instances of the original CP problem. In theory, any metaheuristic technique can be recruited for such a purpose, provided that the reductions carried out in the problem, while seeking to regulate the number of cutting and packing patterns, always respect the constraints imposed by the original problem. This is necessary to ensure that an optimal solution found for any novel problem instance would also be a feasible solution to the original problem. Conversely, the role of the second element, referred to as the Decoder of Reduced Instances (DRI), is to interpret and solve any of the generated problem instances coming out of the GRI. The optimal objective function values achieved with the solving of these sub-problems will serve as a feedback to the DRI, in a manner as to guide its search process. Any exact optimization technique (solver) could be, in thesis, a valid candidate to act as DRI.

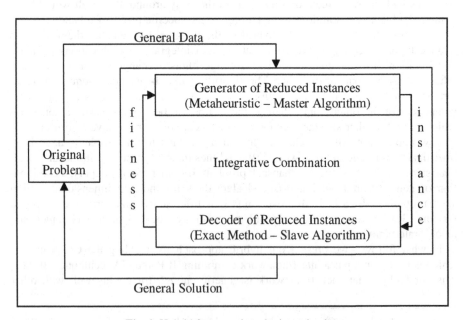

Fig. 1. Hybrid framework under investigation

According to the classification recently proposed by Puchinger and Raidl [12], the methodology falls into the category of integrative combinations. The quality of the solutions to the instances generated by the metaheuristic is determined when the sub-problems are solved by the exact method. The best solution obtained throughout the whole metaheuristic process is deemed as the final solution to the original problem.

Basically, the novel hybrid methodology, as informally described so far, roughly comprehends a sequence of three distinctive steps:

1. **Mathematical formulation of the problem.** In this stage, the aspects to be incorporated in the formal model and the suppositions that can be made relating to the problem itself need to be specified. The desired objectives, the decision variables, and the considered constraints, all need to be made explicit in the mathematical formulation adopted. An erroneous formulation would certainly result in inefficient or incorrect solutions.
2. **Identification of the reducible structures.** In principle, it should be possible to generate all the cutting patterns and solve the CP problem using an exact method. However, the number of cutting patterns tends to grow extraordinarily in practical cases. Thus the idea is to generate only a subset of these patterns in an efficient way. Among the main entities that control the number of cutting patterns are the different types of objects and items, their spatial dimensions, as well as the orientation constraints posed to these items.
3. **Specification of the metaheuristic sub-problem generator.** Finally, the choice of the metaheuristic technique to act as GRI should be done. This issue, like the previous one, deserves greater attention by the framework user, since it is the GRI component that is in charge to discover the reduced version of the original problem that could still provide the more adequate solution to it.

In the following, we provide an instantiation of the proposed hybrid framework to cope specifically with the CL problem. This particular choice was made because, as we have already mentioned in Section 154, the satisfactory tackling of the CL problem comes to be a non-trivial, challenging task, thus serving well the purposes of evaluation of the potentialities behind the novel approach.

2.1 Application to the Container Loading Problem

The three steps of the methodology are revisited in the sequence, each one providing detailed information pertaining exclusively to the CL case considered here. Moreover, although not part of the hybrid framework, a layer construction component was incorporated into it as a particular extension for the CL problem, which is also detailed ahead.

Mathematical Formulation of the CL Problem. To formulate the CL problem, we resort to the ILP model proposed in [8], which was further modified in order to allow the rotation of the boxes. Consider a set of boxes grouped in m types. For each box type i, characterized by its length, width, and height (l_i, w_i, h_i), there is an associated number of boxes b_i. Besides, each box type i presents different modes of orientation in consonance with its orientation constraints. Consider all the n modes originating from the m box types. Each mode j is characterized by its length, width, height, and type (l_j, w_j, h_j, t_j), and has associated with it a given volume v_j. Consider also a container that has L, W, H as its length, width, and height dimensions, respectively, and V as its associated volume. The boxes should be loaded orthogonally into the container. Each 0-1 variable of the model alludes to the decision of whether to place or not a box in mode j at the coordinate (x, y, z). It can be assumed that x, y and z belong to the following discretization sets:

$$X = \left\{ x \middle| x = \sum_{j=1}^{n} \alpha_j l_j, x \le L - l_o, \alpha_j \ge 0, \text{integer} \right\}, \text{where } l_o = \min\{l_j, j = 1,...,n\}, \quad (1)$$

$$Y = \left\{ y \middle| y = \sum_{j=1}^{n} \beta_j w_j, y \le W - w_o, \beta_j \ge 0, \text{integer} \right\}, \text{where } w_o = \min\{w_j, j = 1,...,n\}, \quad (2)$$

$$Z = \left\{ z \middle| z = \sum_{j=1}^{n} \gamma_j h_j, z \le H - h_o, \gamma_j \ge 0, \text{integer} \right\}, \text{where } h_o = \min\{h_j, j = 1,...,n\}. \quad (3)$$

To avoid the interposition of the boxes, the incidence matrix is defined as $g_{jdefpqr}$:

$$g_{jdefpqr} = \begin{cases} 1, \text{if } x_d \le x_p \le x_d + l_j - 1, y_e \le y_q \le y_e + w_j - 1, z_f \le z_r \le z_f + h_j - 1 \\ 0, \text{otherwise} \end{cases}, \quad (4)$$

which should be computed a priori for each mode j ($j = 1, ..., n$), for each position (x_d, y_e, z_f), with $d = 1, ..., |X|$, $e = 1, ..., |Y|$, and $f = 1, ..., |Z|$, and for each position (x_p, y_q, z_r), with $p = 1, ..., |X|$, $q = 1, ..., |Y|$, and $r = 1, ..., |Z|$. Other important parameter definitions come as follows:

$$J_i = \{j | t_j = i\}, \quad (5)$$

$$D_j = \max\{d | x_d \le L - l_j\}, \quad (6)$$

$$E_j = \max\{e | y_e \le W - w_j\}, \quad (7)$$

$$F_j = \max\{f | z_f \le H - h_j\}, \quad (8)$$

with $i = 1,...,m$, $j = 1,...,n$, $d = 1,...,|X|$, $e = 1,...,|Y|$ and $f = 1,...,|Z|$.

Likewise, it is important to consider the specification of the decision variables:

$$a_{jdef} = \begin{cases} 1, \text{if a box in mode } j \text{ is allocated at position } (x_d, y_e, z_f) \\ 0, \text{otherwise} \end{cases}, \quad (9)$$

with $j = 1,...,n$, $d = 1,...,|X|$, $e = 1,...,|Y|$ and $f = 1,...,|Z|$.

Finally, the CL problem can be formulated as:

$$\max \sum_{j=1}^{n} \sum_{d=1}^{D_j} \sum_{e=1}^{E_j} \sum_{f=1}^{F_j} \frac{v_j}{V} \times a_{jdef}$$

$$s.a. \ \sum_{j=1}^{n} \sum_{d=1}^{D_j} \sum_{e=1}^{E_j} \sum_{f=1}^{F_j} g_{jdefpqr} a_{jdef} \leq 1, p = 1,...,|X|, q = 1,...,|Y|, r = 1,...,|Z|$$

$$\sum_{j \in J_i} \sum_{d=1}^{D_j} \sum_{e=1}^{E_j} \sum_{f=1}^{F_j} a_{jdef} \leq b_i, i = 1,...,m$$

with $a_{jdef} \in \{0,1\}$

(10)

Identification of the Reducible Structures. The number of packing patterns of the model proposed for the CL problem under investigation can be defined by the amount of box modes n and by the cardinality of the discretization sets, namely $|X|$, $|Y|$, and $|Z|$. Several reduced instances of the original problem can then be eventually generated by excluding some of the box modes and/or some positions demarcated by the discretization sets.

Specification of the Metaheuristic Sub-problem Generator. A modified genetic algorithm has been implemented to account for the GRI task. Each of its individuals is represented by a binary chromosome, which encodes a particular set of discretization points to the aforementioned ILP model. In this encoding, each gene (bit) represents one possible element of discretization along a particular dimension, as shown in Fig. 2. Assigning a null value to a bit means that its associated element of discretization will not be used for the construction of the reduced problem; by flipping the value of this gene to one, the element will take part in the reduced CL problem instance.

As mentioned earlier, each sub-problem should be solved by an exact method playing the role of the DRI. The optimal solution value attained for this problem instance – which represents the actual space used up in the container – is assigned as

Discretization Sets:

$X = \{0, 3, 5, 6, 8, 9\}$ $\rightarrow$ $x_1 x_2 x_3 x_4 x_5 x_6$
$Y = \{0, 4, 7, 8\}$ $\rightarrow$ $y_1 y_2 y_3 y_4$
$Z = \{0, 2, 4, 6, 7, 8\}$ $\rightarrow$ $z_1 z_2 z_3 z_4 z_5 z_6$

Chromosome:

| x_1 | x_2 | x_3 | x_4 | x_5 | x_6 | y_1 | y_2 | y_3 | y_4 | z_1 | z_2 | z_3 | z_4 | z_5 | z_6 |
| 1 | 0 | 0 | 1 | 0 | 1 | 1 | 1 | 0 | 1 | 1 | 1 | 1 | 0 | 1 | 0 |

Fig. 2. The chromosome representation

the fitness value of its corresponding GA individual. For those cases where it is not possible to solve, or even generate, the instance model, due to the sub-problem's inherent complexity and/or to other computational limitations of the solver environment, the corresponding GA individual is discarded from the pool and another chromosome is randomly generated to take its place.

For the initial population, p chromosomes are generated in the following manner: One of the modes is randomly chosen and, then, only the elements proportional to the respective dimensions of this mode will be considered for being selected to take part in the problem instance. Standard selection/variation operators are applied iteratively over the individuals across generations. More precisely, individuals are selected for mating and for replacement with the help of the roulette wheel selection mechanism. Furthermore, the single-point crossover is adopted with rate τ_c, whereas the simple mutation operator is applied with different rates for each chromosome segment. These mutation rates vary according to the length of the respective discretization set: $\tau_{mX} = 1/|X|$, $\tau_{mY} = 1/|Y|$, and $\tau_{mZ} = 1/|Z|$. A maximum number of iterations τ_g, a prefixed interval of convergence c, as well as a maximum limit on the execution time of the whole framework t (in hours), are used in concert as stopping criteria.

Layer Construction Heuristics. To cope specifically with the CL problem, a layer construction component was incorporated to limit the effective size of the container. By this means, each generated layer can be treated as a distinct CL problem, which must be solved by the hybrid algorithm. The solution of the original problem is constructed in sequential steps. Two strategies were used to control the effective length of the layer: (a) *MaxMax*, whereby the size of the container is set in accordance with that type of box with the largest dimension; and (b) *MaxMin*, whereby the size is regulated by that type of box showing the largest of the overall small dimensions. When the size of the free length of the container becomes smaller than the size of the layer, the remaining space is considered in this iteration.

3   Computational Experiments

To evaluate the potentialities behind the novel approach, a series of experiments have been conducted by our group, and some of the preliminary results achieved so far are discussed in the following subsections. For conducting these experiments, the GRI metaheuristic (GA) was developed in *Delphi* language, whereas an implementation of the overall hybrid framework has been executed and tested on an Intel Pentium 4 3.2 GHz platform with 1GB of RAM memory available. For implementing the DRI component of the framework, we have made use of a *LINGO 8.0 DLL* as well.

3.1   Benchmarking Problems

Some experiments have been performed over the Thpack group of problems of the OR-Library available at http://people.brunel.ac.uk/~mastjjb/jeb/info.html. For these problems, the two layer-construction strategies described in the previous section, viz. *MaxMax* and *MaxMin*, have been adopted in turn. (Actually, for each strategy, only one separate execution of the framework is carried out.) The meta-parameters of the

GA engine have been set as follows, after some experimental calibration: $p = 50$ individuals, $\tau_c = 0.5$, $\tau_g = 50$ iterations, $c = 10$ iterations, and $t = 1$ hour.

In Table 1, the best result – in terms of volume utilization of the container V_{HA} – achieved by our hybrid approach (HA) for each problem of the Thpack1 group is shown, along with the heuristic strategy adopted. Conversely, in Tables 2 and 3, we contrast the performance of the hybrid approach with that exhibited by the well-known combined heuristic B/R, proposed by Bischoff and Ratcliff [3], in those problem instances where the latter have presented the worst and best performance indices, respectively. These instances are presented in accordance with the amount of box types considered (groups Thpack1–Thpack7).

Overall, our hybrid methodology has been obtaining satisfactory results with these benchmarking problems. For instance, on average, considering all the problem instances in Table 1, we have achieved the mark of 87.52% (±3.49) of effective usage of the container space, outperforming the 85.40% (±4.30) score achieved by the combined heuristic B/R. Considering the results in Tables 2 and 3, the hybrid algorithm outperforms the heuristic B/R in the worst-case instances, whereas, in the best-case instances, the heuristic B/R prevails. On average, the new methodology has shown superior performance, reaching the mark of 86.53% of effective volume utilization of the container across all problem instances against the 83.53% score achieved by the heuristic B/R.

Table 1. Results obtained for the first 50 problems of the Thpack1 group

Problem	V_{HA} (%)	Strategy	Problem	V_{HA} (%)	Strategy
01	89.29	MaxMax	26	83.19	MaxMax
02	89.95	MaxMin	27	86.00	MaxMax
03	81.02	MaxMin	28	88.51	MaxMin
04	88.50	MaxMin	29	91.72	MaxMax
05	90.26	MaxMax	30	90.08	MaxMin
06	90.11	MaxMax	31	89.17	MaxMin
07	81.41	MaxMax	32	92.06	MaxMax
08	88.22	MaxMax	33	84.36	MaxMin
09	87.60	MaxMax	34	89.40	MaxMax
10	87.92	MaxMax	35	84.04	MaxMin
11	89.19	MaxMin	36	89.11	MaxMax
12	88.18	MaxMin	37	91.75	MaxMax
13	90.11	MaxMax	38	80.85	MaxMax
14	86.37	MaxMin	39	89.15	MaxMax
15	85.92	MaxMax	40	88.79	MaxMax
16	86.98	MaxMax	41	86.44	MaxMin
17	90.64	MaxMin	42	91.34	MaxMin
18	80.55	MaxMax	43	91.37	MaxMax
19	90.37	MaxMin	44	76.01	MaxMin
20	90.57	MaxMin	45	88.80	MaxMax
21	89.27	MaxMin	46	87.73	MaxMax
22	87.89	MaxMax	47	82.16	MaxMax
23	83.03	MaxMax	48	88.17	MaxMax
24	86.35	MaxMin	49	89.12	MaxMax
25	87.31	MaxMax	50	91.23	MaxMax

Table 2. Comparative results for the worst case instances of B/R in groups Thpack1–Thpack7

Group	Problem	B/R (%)	V_{HA} (%)	Strategy
Thpack1	44	73.72	76.01	MaxMin
Thpack2	11	73.79	86.19	MaxMax
Thpack3	46	75.33	87.88	MaxMax
Thpack4	93	78.38	88.10	MaxMin
Thpack5	70	78.71	86.94	MaxMin
Thpack6	93	75.22	87.60	MaxMin
Thpack7	74	75.73	87.89	MaxMin

Table 3. Comparative results for the best case instances of B/R in groups Thpack1–Thpack7

Group	Problem	B/R (%)	V_{HA} (%)	Strategy
Thpack1	39	94.36	89.15	MaxMax
Thpack2	33	93.76	87.47	MaxMin
Thpack3	49	92.63	90.54	MaxMin
Thpack4	22	90.06	85.26	MaxMin
Thpack5	41	90.39	87.18	MaxMin
Thpack6	58	89.15	86.15	MaxMax
Thpack7	51	88.28	85.11	MaxMin

3.2 Case Study

The hybrid methodology has also been applied in the ambit of a large metallurgic factory. This particular industrial unit has usually made use of a load planning and accommodation system that is quite well known in the market, namely *MaxLoad Pro* [14]. In Table 4, some configuration data related to the problems being tackled by our group are presented, where L, W, and H refer to the length, width, and height dimensions of the container, respectively, while l, w, and h refer to the length, width, and height dimensions of the boxes in that order, and b denotes the amount of items demanded in the order.

In Table 5, some results are displayed contrasting the performance achieved by the proposed hybrid approach (HA) with that achieved by *MaxLoad* (ML). In this table, X_{ML} and V_{ML} refer to the number of boxes allocated and the volume effectively used in the container produced with the ML system, respectively, while X_{HA} and V_{HA} refer to the number of boxes allocated and the space effectively occupied in the container produced with the help of the hybrid methodology.

Table 4. Configuration data for the case study

Problem	L	W	H	l	w	h	b
01	12000	2340	2680	1585	655	580	120
02	12000	2340	2680	1730	650	640	80
03	12000	2340	2680	870	585	525	280
04	12000	2340	2680	870	585	785	180
05	5810	2330	2380	870	585	525	120
06	5810	2330	2380	1585	655	580	50

Table 5. Results obtained for the practical cases studied

Problem	X_{ML}	V_{ML} (%)	X_{HA}	V_{HA} (%)
01	101	80,8	117	93,2
02	80	76,5	80	76,5
03	264	93,7	272	96,6
04	180	95,6	180	95,6
05	110	91,2	115	95,4
06	45	84,1	47	87,8

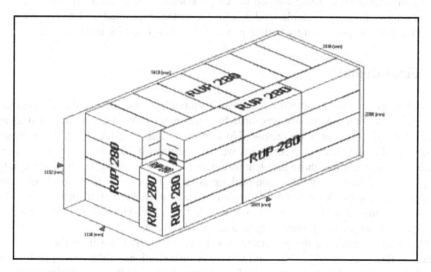

Fig. 3. Packing pattern produced by ML, with 45 loaded boxes (instance #06)

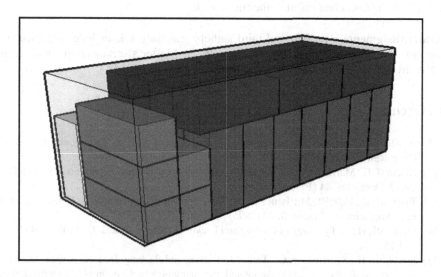

Fig. 4. Packing pattern produced by HA, with 47 loaded boxes (instance #06)

Among the six problem instances considered, the hybrid approach was able to find the best solutions in four of the cases, with a tie in the other two. The most expressive result was obtained in instance #01, where the difference between the two contestants, in terms of the volume effectively used in the container, was of 12.4%. It is worth to emphasize that, in this circumstance, the hybrid methodology was applied with no layer construction heuristic within. Moreover, the control parameters of the GA engine were set as follows: $p = 50$ individuals, $\tau_c = 0.5$, $\tau_g = 50$ iterations, $c = 10$ iterations, and $t = 5$ hours. Although the current instantiation of the framework has not yet explicitly considered constraints on the cargo stability, we could verify that the solutions achieved represent stable loadings for the instances studied. In Figs. 3 and 4, the packing patterns produced by both approaches are illustrated for instance #06.

4   Final Remarks

In this work, we have introduced a novel methodology for solving the CL problem, which is based on a particular type of hybridization between an exact (ILP) and an approximate (GA) optimization method. Up to now, the results achieved by the approach, with regard to the series of experiments we have performed, have been very satisfactory, taking into account that those problem instances could not be directly solved through the application of the exact optimization package alone. For both scenarios considered in this paper, viz. benchmarking problems and a real case study, the hybrid methodology has presented competitive solutions to those provided by a well-known heuristic and a commercial tool, respectively, prevailing on average over them.

In order to evaluate the generalization level behind the novel methodology, we plan, as future work, to develop and test other instantiations of the conceptual framework for coping with other types of CP problems. Likewise, it is in our plans to investigate other types of metaheuristics (like trajectory-based or even hybrid ones) to play the role of the GRI component in the framework.

Acknowledgments. The first and third authors gratefully acknowledge the financial support received from FUNCAP, through a Master's scholarship, and from CNPq/FUNCAP, through a DCR grant (project number 23661-04).

References

[1] Beasley, J.: An Exact Two-dimensional Non-guillotine Cutting Tree Search Procedure. Oper. Res. **33** (1985) 49-64

[2] Bischoff, E., Marriot, M.: A Comparative Evaluation of Heuristics for Container Loading. Eur. J. Oper. Res. **44** (1990) 267-276

[3] Bischoff, E., Ratcliff, M.: Issues in the Development of Approaches to Container Loading. Omega Int. J. Manage. S. **23** (1995) 377-390

[4] Dyckhoff, H.: A Typology of Cutting and Packing Problems. Eur. J. Oper. Res. **44** (1990) 145-159

[5] Dyckhoff, H., Scheithauer, G., Terno, J.: Cutting and Packing. In: Dell'Amico, M., Maffioli, F., Martello, S. (eds.): Annotated Bibliographies in Combinatorial Optimization, John Wiley & Sons, New York (1997) 393-412

[6] Dumitrescu, I., Stützle, T.: Combinations of Local Search and Exact Algorithms. In: Cagnoni et al (eds.): Applications of Evolutionary Computing. Lecture Notes in Computer Science **2611**. Springer-Verlag, Berlin Heidelberg New York (2003) 211-223

[7] Gilmore, P., Gomory, R.: A Linear Programming Approach to the Cutting Stock Problem (Part I). Oper. Res. **9** (1961) 849-859

[8] Morabito, R., Arenales, S.: Abordagens para o Problema do Carregamento de Contêineres. Pesquisa Operacional **17** (1997) 29-56 (in Portuguese)

[9] Plateau, A., Tachat, D., Tolla, P.: A Hybrid Search Combining Interior Point Methods and Metaheuristics for 0-1 Programming. Int. T. Oper. Res. **9** (2002) 731-746

[10] Puchinger, J.: Combining Metaheuristics and Integer Programming for Solving Cutting and Packing Problems. PhD thesis, Vienna University of Technology, Institute of Computer Graphics and Algorithms (2006)

[11] Puchinger, J., Raidl, G., Koller, G.: Solving a Real-World Glass Cutting Problem. In: Gottlieb, J., Raidl, G. (eds.): Evolutionary Computation in Combinatorial Optimization, Lecture Notes in Computer Science **3004**. Springer-Verlag, Berlin Heidelberg New York (2004) 162-173

[12] Puchinger, J., Raidl, G.: Combining Metaheuristics and Exact Algorithms in Combinatorial Optimization: A Survey and Classification. In: Mira, J., Álvarez, J. (eds.): Interplay Between Natural and Artificial Computation, Lecture Notes in Computer Science **3562**. Springer-Verlag, Berlin Heidelberg New York (2005) 41-53

[13] Terno, J., Scheithauer, J., Sommerweiß, U.: An Efficient Approach for the Multi-pallet Loading Problem. Eur. J. Oper. Res. **123** (2000) 372-381

[14] TOPS Engineering Corporation. MaxLoad Pro Version 2.7X User's Guide (2004).

[15] Yeung, L. H. W., Tang, W. K. S.: A Hybrid Genetic Approach for Container Loading in Logistics Industry. IEEE T. Ind. Electron. **52** (2005), 617-627

A Population-Based Local Search for Solving a Bi-objective Vehicle Routing Problem

Joseph M. Pasia[1,2], Karl F. Doerner[2], Richard F. Hartl[2], and Marc Reimann[3]

[1] Department of Mathematics, University of the Philippines-Diliman,
Quezon City, Philippines
jmpasia@up.edu.ph
[2] Department of Management Science, University of Vienna,
Vienna, Austria
{richard.hartl, karl.doerner}@univie.ac.at
[3] Institute for Operations Research, ETH Zurich,
Zurich, Switzerland
marc.reimann@ifor.math.ethz.ch

Abstract. In this paper we present a population-based local search for solving a bi-objective vehicle routing problem. The objectives of the problem are minimization of the tour length and balancing the routes. The algorithm repeatedly generates a pool of good initial solutions by using a randomized savings algorithm followed by local search. The local search uses three neighborhood structures and evaluates the fitness of candidate solutions using dominance relation. Several test instances are used to assess the performance of the new approach. Computational results show that the population-based local search outperforms the best known algorithm for this problem.

1 Introduction

The vehicle routing problem (VRP), introduced in [1], consists of finding the optimal route for a fleet of vehicles, starting and ending at a single depot, that must serve a set of n customer demands such that each customer is visited by only one vehicle route. If each vehicle can only collect a maximum capacity of Q units of demands, then the problem is known as capacitated VRP (CVRP).

Although exact approaches [2,3] have been proposed to solve the VRP, many approximation methods have been developed since VRP has been proven $\mathcal{N}$P-hard [4]. Some examples of these methods are the classical heuristics such as the well-known savings algorithm [5]. These methods put more emphasis on the ability to search a good feasible solution.

Over the last 15 years, increasing research effort has been devoted on the development of metaheuristic approaches since they can search the solution space much more thoroughly. The metaheuristic approaches use either the principle of local search or population search [6,7]. Local search methods approximate a region of the Pareto front in the direction given by a weight vector λ using a single solution. The aggregation of the objectives via λ is usually used in order

C. Cotta and J. van Hemert (Eds.): EvoCOP 2007, LNCS 4446, pp. 166–175, 2007.

to focus the search on the region of interest. On the other hand, population searches such as the adaptive memory procedure [8] and SavingsAnts [9] maintain a pool of solutions called population that move together through evolution process towards the Pareto front. In many cases, local search methods are used to improve the quality of the solutions in the population.

In recent years, metaheuristics have also been used to solve vehicle routing problems with multiple objectives. For instance, a bi-objective CVRP with route balancing (CVRPRB) was tackled using genetic algorithms [10,11]. The two objectives of CVRPRB are (i) to minimize the sum of the total distance travelled by each vehicle and (ii) to minimize the difference between the longest and shortest vehicle tours.

In general, there is no single solution that simultaneously accomplishes the objectives of a bi-objective optimization problem. Hence, the Pareto optimal solutions or sometimes called the set of efficient solutions are considered. We say that a solution x is an efficient solution if there exists no other feasible solution y such that $f_k(y) \leq f_k(x)$, for $k = 1, 2$ and $f_k(y) < f_k(x)$ for some k. Otherwise, we say that x is dominated by y and we denote this by $y \prec x$. In addition, if x^* is Pareto optimal then $z^* = f(x^*)$ is called nondominated vector and the set of all nondominated vectors is referred to as nondominated frontier (or Pareto front or trade-off surface).

In this paper, we propose another approach in solving the CVRPRB by combining the principles of population search and local search. Our approach starts by generating a pool of good starting solutions using a randomized savings algorithm. These solutions then undergo local search that uses three neighborhood structures. The candidate solutions in the local search are evaluated using dominance relation. In the following discussions, we will refer to our method as population-based local search or P-LS.

This paper is organized as follows: Section 2 explains the details of P-LS. Section 3 discusses the numerical results of the study and Section 4 provides a short conclusion of the study.

2   Population-Based Local Search for CVRPRB

The P-LS method follows the two basic steps given by Algorithm 1. The first step is the initialization phase or the creation of the starting solutions. The second step is the local search phase where we apply our local search operators. In the following sections, we describe the details how we implemented these steps.

Algorithm 1. Basic algorithmic framework of P-LS

While (condition is satisfied) Do /*We call this loop as generation*/
    Create a pool of solutions $\mathcal{S}$
    Apply local search on the solutions of $\mathcal{S}$

2.1   Initialization Phase of P-LS

Being a population-based heuristic, the starting solutions of P-LS are important. It was demonstrated in [12,13,14] that spending time in creating a good initial population improves the convergence in optimization. Hence, we use a pool of good starting solutions for our P-LS. Our strategy in creating the initial solutions uses the randomized savings algorithm or savings algorithm with candidate list. The savings algorithm starts with the assignment of each customer to a separate tour. The customers or the partial tours are then combined based on the savings values given by $s(i,j) = d(i,0) + d(j,0) - d(i,j)$ where $d(i,j)$ is the distance between customers i and j and the index 0 denotes the depot. Clearly, $s(i,j)$ is the value saved when i and j are combined instead of serving them by two different tours. The combination of customers begins with the largest savings until no more combination is feasible.

Instead of combining the customers having the largest savings, the randomized savings algorithm creates a candidate list C of feasible combinations of customers i and j. The set C consists of the combinations with the $|C|$ best savings values. Each combination in the candidate list is selected with equal probability. After combining the selected pair of customers, the combinations in C that become infeasible are removed and we maintain the size of C by adding the next feasible combinations with the best savings values.

To generate the starting solutions of P-LS, a pool S of identical solutions are initially created. These identical solutions assign each customer to a separate tour. We improve each of these solutions by combining the customers based on the randomized savings algorithm. We then apply 2-opt local operator to the solutions of S in order to avoid artificially balanced solutions [11]. In P-LS, we apply our local search operators only to the nondominated solutions. Hence, we remove all dominated solutions in S before proceeding to the local search phase.

2.2   Local Search Phase of P-LS

The local search phase uses one intratour-neighborhood called 2-opt and two intertour-neighborhood structures namely move and swap. The move neigborhood inserts the customer of one partial tour to another partial tour. On the other hand, the swap neighborhood exchanges two customers from different partial tours. After performing either the insertion or exchange operator, the 2-opt operator is applied to the affected partial tours.

The basic step of the local search phase is described in Algorithm 2. This step was called multiobjective local search in [15] and pareto local search in [16] and it was implemented in scheduling problem using a single neighborhood structure. The main difference of our approach compared to the existing techniques is that we use three different neighborhood structures. It has been shown in [9,17] that these three neighborhoods are effective for the single objective classical VRP.

Starting from a solution $z \in S$, the feasible solutions of the first neighborhood $\mathcal{N}_1$ of z are explored. All feasible neighboring solutions are compared and the dominated ones are removed. Each of the remaining efficient solution will undergo the same process as z i.e., its entire neighborhood is searched and all the dominated solutions are removed. We repeat the entire process of searching the whole neighborhood and removing the dominated solutions until all solutions in the neighborhood are dominated. When this happen, Algorithm 2 is repeated on $\mathcal{P}$ using the second neighborhood $\mathcal{N}_2$. Unless the termination conditions are not satisfied after performing $\mathcal{N}_2$, we apply $\mathcal{N}_1$ again on the pareto set returned by $\mathcal{N}_2$. Since exploring the entire neighborhood of a given solution is computationally expensive, we only allow a certain number of solutions to undergo the local search. Figure 1 illustrates this process.

Algorithm 2. Basic step of the local search phase

Pareto set $\mathcal{S}$
t = 0
While ($\mathcal{S}$ is non-empty and t < M) Do /*We call this loop as iteration*/
   Pareto set $\mathcal{C} = \emptyset$
   Forall $z \in \mathcal{S}$ do
      Pareto set $\mathcal{L} \leftarrow z$
      Forall $w \in \mathcal{N}_i(z)$ do
         If $w \notin \mathcal{L}$ and w is not dominated by $\mathcal{L}$ then
         Set $\mathcal{L}$ = nondominated solutions of $(\mathcal{L} \cup \{w\})$ and
         Set $\mathcal{C}$ = nondominated solutions of $(\mathcal{C} \cup \{w\})$
   $\mathcal{S} = \mathcal{C}$
   Update Pareto set $\mathcal{P}$ by $\mathcal{C}$
   t = t + 1
Return $\mathcal{P}$

---

3   Numerical Results

The numerical analysis was performed on a set of benchmarks described in [18]. The set of benchmarks consists of 7 test instances having 50 to 199 customers and a single depot. Four of the instances were generated such that the customers are uniformly distributed on a map and the remaining instances feature clusters of customer locations. All test instances have capacity constraints. We performed all our methods on a personal computer with 3.2 Ghz processor; the algorithms were coded in C++ and compiled using GCC 4.1.0 compiler.

3.1   Evaluation Metrics

The use of unary quality indicators has become one of the standard approaches in assessing the performance of different algorithms for bi-objective problems. It complements the traditional approach of using graphical visualization which may provide information on how the algorithm works [19]. This study considered three unary quantitative measures namely, the *hypervolume indicator*, *unary epsilon indicator*, and *R3 indicator*.

Th hypervolume indicator (I_H) measures the hypervolume of the objective space that is weakly dominated by an approximation set or the set containing the

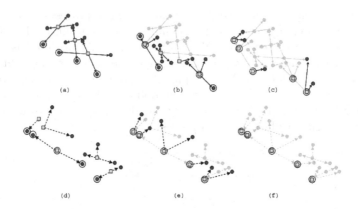

Fig. 1. The boxes represent the efficient solutions. The solid circles are the feasible solutions in the neighborhood of an efficient solution. The encircled solutions are the efficient neighbors which become the new efficient solutions of $\mathcal{S}$. (a) Four efficient solutions are found using $\mathcal{N}_1$. (b) After exploring the neighborhoods of these efficient solutions, 3 new efficient solutions are found and 2 efficient solutions from (a) remain efficient. (c) Not a single neighbor of efficient solutions in (b) is efficient. (d) $\mathcal{N}_2$ is then used and 4 of the neighbors are found to be efficient. (e) The $\mathcal{N}_2$-neighborhood of the efficient solutions in (d) does not have any efficient solution this time. (f) Since $\mathcal{N}_2$ generated new efficient solutions, the current set of efficient solutions will again be explored using $\mathcal{N}_1$.

nondominated frontier of an approximation method [20]. This is calculated using a boundary point that is dominated by all approximation sets. It has a desirable property that whenever an approximation set A is better than approximation set B, then the hypervolume of A is greater than B.

The unary epsilon indicator I_ϵ gives the minimum factor ϵ such that if every point in reference set X is multiplied by ϵ, then the resulting approximation set is weakly dominated by A. For minimization problem, this indicator is formally defined by:

$$I_\epsilon(A) = I_\epsilon(A, X) = \inf_{\epsilon \in X} \left\{ \forall z^2 \in X \exists z^1 \in A : z^1 \preceq_\epsilon z^2 \right\} \qquad (1)$$

where the ϵ-dominance relation is defined as $z^1 \preceq_\epsilon z^2 \Leftrightarrow \forall i \in 1, 2, \ldots, n : z_i^1 \leq \epsilon \cdot z_i^2$. Note that a small ϵ value is preferable.

The R3 indicator (I_{R3}) used in this study is one of R indicators proposed in [21]. Given a set of weight vectors Λ, this indicator is defined as:

$$I_{R3}(A) = I_{R3}(A, X) = \frac{\sum_{\lambda \in \Lambda} \left[u^*(\lambda, X) - u^*(\lambda, A) \right] / u^*(\lambda, A)}{|\Lambda|} \qquad (2)$$

where u^* is the maximum value attained by a utility function u_λ with weight λ, i.e., $u^*(\lambda, A) = \max_{z \in A} u_\lambda(z)$. In this study, the utility function is given by:

$$u_\lambda(z) = -\left(\max_{j\in 1..n} \lambda_j |z_j^* - z_j| + \rho \cdot \sum_{j\in 1..n} |z_j^* - z_j|\right) \qquad (3)$$

where z^* is the ideal point and ρ is a sufficiently small positive real number. The values of I_{R3} range from -1 to 1 where values close to -1 are superior.

3.2   Parameter Settings

The initial number of solutions in each pool S is equal to the number of customers. The size of the candidate list is given by $\lfloor 0.10 \times (\# \text{ of customers}+1)\rfloor$. The maximum number of solutions in S that undergo local search is ten. These 10 solutions are chosen so that the corresponding nondominated points are evenly distributed in the objective space i.e., dividing the relevant section in equally-sized segments. The value of M is either 100 or 150 for the move-neighborhod and 50 to 75 for the swap-neighborhood.

3.3   Analysis

Ten runs with different random seeds were performed for each of the test instance. Before applying the different unary indicators, all approximation sets are normalized between 1 and 2. The boundary point used in the hypervolume indicator is (2.1, 2.1), and the ρ and $|\Lambda|$ in I_{R3} are 0.01 and 500 respectively [19]. The reference set for each test instance consists of the points that are not dominated by any of the approximation sets generated by all algorithms under consideration.

The performance of P-LS is compared to the Nondominated Sorting Genetic Algorithm II (NSGA II) developed in [22] and proposed as algorithm for CVR-PRB in [10]. The NSGA II approach for CVRPRB was enhanced by parallelization and by the use of elitist diversification mechanism. Several implementations of NSGA II which have different number of Power4 1.1 Ghz processors were examined. The average computation time ranged from 900 to 5200 seconds. We used the only available results of one of the NSGA II variants found in http://www2.lifl.fr/jozef/results_VRPRB.html for comparison.

In this study, we demonstrate the importance of randomized savings algorithm in generating the pools of initial solutions. To do this, we compare our P-LS to a P-LS that generates solutions randomly. We call this method as P-LS-0. In addition, we also investigate the advantage of allowing our P-LS to use more starting solutions by comparing our method to a P-LS where M=∞. We refer to this method as P-LS-1. All three methods are terminated after reaching a computational time that ranges from 600 to 30000 seconds depending on the test instance.

Figure 2 shows the boxplots of the different unary indicators for all test instances[1]. A boxplot provides a graphical summary of the median, the range and

---

[1] The first number in the name of each test instance corresponds to the total number of customers and depot. The last letter determines whether the customers are clustered (c) or uniformly distributed (e).

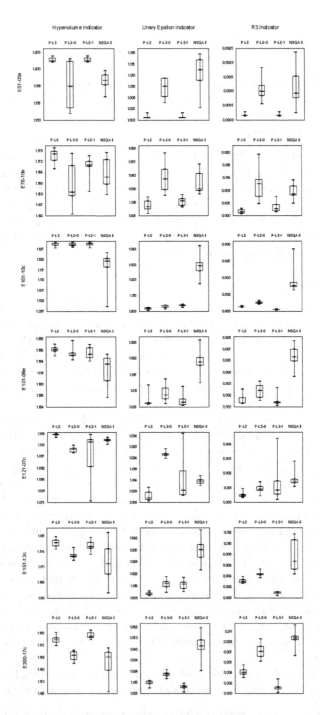

Fig. 2. Box plots of the different unary indicators for all test instances. High hypervolume values and low unary epsilon and R3 values are preferable.

inter-quartile range, and the orientation of the median relative to the quartiles for a set of data.

P-LS and P-LS-0. For all test instances, the boxplots of P-LS are better than that of P-LS-0. It can be observed that the positive effect of the randomized savings algorithm is more evident in larger instances (at least 120 customers). Thus the good performance of P-LS can be attributed to the quality of starting solutions generated by the randomized savings algorithm.

P-LS and P-LS-1. Allowing the P-LS to have more pools improves the performance P-LS with respect to hypervolume and unary epsilon indicators in all but one test instance. However, the R3 indicator gets worse in many test instances. To explain why this may happen, we consider the first runs of P-LS and P-LS-1 for test instance E151-12c. Figure 3 provides the unary quality indicators and the plots of the efficient points for this case. It is clear from these plots that the best objective values of P-LS with respect to total distance are much better than that of P-LS-1 and this translates into bigger hypervolume. In addition, P-LS-1 is slightly better than P-LS in the middle region. It has more points in this region and most of them belong to the reference set. This slight difference in the middle area may not compensate the gains of P-LS with respect to hypervolume but this helps P-LS-1 to have a better R3 indicator value. Regarding the unary epsilon, it is also clear from the graph that P-LS-1 requires a bigger factor in order for its approximation set to weakly dominate the reference set.

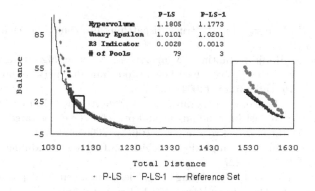

Fig. 3. Plots the nondominated frontiers of the first run of P-LS and P-LS-1 for test instance E151-12c. The region in the middle enclosed in a box is enlarged.

P-LS and NSGA II. The boxplots of the three unary quality indicators of P-LS is better than NSGA II for all instances. In fact, all the median values of P-LS are better than the median values of NSGA II and for many cases, the median values are better than the best values of NSGA II. This shows that our P-LS approach performs well in all the test instances that were used. It is also interesting to note that P-LS-0 and P-LS-1 are also better than NSGA II with respect to the three unary quality indicators for some instances. For example, all three P-LS approaches are better than NSGA II in test instances E101-10c and E101-08e.

4 Conclusion

In this study, we proposed the P-LS or population-based local search to solve a bi-objective vehicle routing problem. The P-LS uses a pool of good starting solutions generated by a randomized savings algorithm. All efficient solutions in the pool were allowed to undergo local search. Our local search approach uses three neighborhoods and and the candidate solutions were evaluated using dominance relation.

Computational results showed that our P-LS approach performed well compared to NSGA II with respect to three unary quality indicators. We have also demonstrated that using the randomized savings algorithm improves the solutions quality of P-LS. We have also shown that the performance of P-LS improves when we allow our P-LS to use more starting pools of initial solutions.

In the future, we intend to apply our P-LS to the large-scale VRP instances. We also intend to extend our P-LS approach to VRP with more than two objectives, e.g. load balancing, number of vehicles. Moreover, we plan to apply P-LS to other multiobjective combinatorial problems.

References

1. Dantzig, G., Ramsey, J.: The truck dispatching problem. Management Science **6** (1959) 80–91
2. Baldacci, R., Hadjiconstantinou, Mingozzi, A.: An exact algorithm for the capacitated vehicle routing problem based on a two-commodity network flow formulation. Operations Research **52**(5) (2004) 723–738
3. Fukasawa, R., Longo, H., Lysgaard, J., Poggi de Aragão, M., Reis, M., Uchoa, E., Werneck, R.: Robust branch-and-cut-and-price for the capacitated vehicle routing problem. Mathematical Programming **106**(3) (2006) 491–511
4. Lenstra, J., Kan, A.: Complexity of vehicle routing and scheduling problem. Networks **11** (1981) 221–227
5. Clarke, G., Wright, J.: Scheduling of vehicles from a central depot to a number of delivery points. Operations Research **12** (1964) 568–581
6. Cordeau, J., Gendreau, M., Laporte, G., Potvin, J., Semet, F.: A guide to vehicle routing heuristics. Journal of the Operational Research Society **53** (2002) 512–522
7. Ehrgott, M., Gandibleux, X.: A survey and annotated bibliography of multiobjective combinatorial optimization. OR Spektrum **22** (2000) 425–460
8. Rochat, Y., Taillard, E.: Probabilistic diversification and intensification in local search for vehicle routing. Journal of Heuristics **1** (1995) 147–167
9. Doerner, K., Gronalt, M., Hartl, R.F., Reimann, M., Strauss, C., Stummer, M.: SavingsAnts for the vehicle routing problem. In Cagnoni, S., et al., eds.: Proceedings of Applications of Evolutionary Computing : EvoWorkshops 2002. Volume 2279 of LNCS. (2002) 11–20

10. Jozefowiez, N., Semet, F., Talbi, E.G.: Enhancements of NSGA-II and its applica-
    tion to the vehicle routing problem with route balancing. In Talbi, E., et al., eds.:
    Proceedings of 7th International Conference Artificial Evolution-EA 2005. Number
    3871 in LNCS, Springer-Verlag (2006) 131–142
11. Jozefowiez, N., Semet, F., Talbi, E.G.: Parallel and hybrid models for multi-
    objective optimization: Application to the vehicle routing problem. In Guervós,
    J., et al., eds.: Parallel Problem Solving from Nature—PPSN VII. Number 2439 in
    LNCS, Granada, Spain, Springer-Verlag (2002) 271–280
12. Haubelt, C., Gamenik, J., Teich, J.: Initial population construction for convergence
    improvement of moeas. In Coello, C., Aguirre, A., Zitzler, E., eds.: Evolutionary
    Multi-criterion Optimization, EMO'2005. Volume 3410 of LNCS. (2005) 191–205
13. Gandibleux, X., Morita, H., Katoh, N.: The supported solutions used as a genetic
    information in a population heuristic. In: Evolutionary Multi-criterion Optimiza-
    tion, EMO'2001. Volume 1993 of LNCS., Zurich, Switzerland (2001) 429–442
14. Morita, H., Gandibleux, X., Katoh, N.: Experimental feedback on biobjective
    permutation scheduling problems solved with a population heuristic. Foundations
    of Computing and Decision Sciences 26(1) (2001) 23–50
15. Arroyo, J., Armentano, V.: Genetic local search for multi-objective flowshop schedul-
    ing problems. European Journal of Operational Research 167 (2005) 717–738
16. Basseur, M., Seynhaeve, F., Talbi, E.: Path relinking in pareto multi-objective
    genetic algorithms. In Coello, C., Aguirre, A., Zitzler, E., eds.: Evolutionary Multi-
    criterion Optimization, EMO'2005. Volume 3410 of LNCS. (2005) 120–134
17. Reimann, M., Doerner, K., Hartl, R.: D-ants: Savings based ants divide and con-
    quer the vehicle routing problem. Computers & Operations Research 31(4) (2004)
    563–591
18. Christofides, N., Mingozzi, A., Toth, P.:   The vehicle routing problem.   In
    Christofides, N., Mingozzi, A., Toth, P., Sandi, C., eds.: Cominbatorial Optimiza-
    tion. John Wiley and Sons (1979)
19. Knowles, J., Thiele, L., Zitzler, E.: A tutorial on the performance assessment
    of stochastic multiobjective optimizers. Technical Report TIK-Report No. 214,
    Computer Engineering and Networks Laboratory, ETH Zurich, Gloriastrasse 35,
    ETH-Zentrum, 8092 Zurich, Switzerland (2006)
20. Zitzler, E., Thiele, L.: Multiobjective evolutionary algorithms: a comparative case
    study and the strength pareto approach. IEEE Trans. Evolutionary Computation
    3(4) (1999) 257–271
21. Hansen, M., Jaszkiewicz, A.: Evaluating the quality of approximations to the non-
    dominated set. Technical Report Technical Report IMM-REP-1998-7, Technical
    University of Denmark (1998)
22. Deb, K., Agrawal, S., Pratap, A., Meyarivan, T.: A fast elitist non-dominated
    sorting genetic algorithm for multi-objective optimisation: NSGA-II. In: PPSN.
    (2000) 849–858

Combining Lagrangian Decomposition with an Evolutionary Algorithm for the Knapsack Constrained Maximum Spanning Tree Problem

Sandro Pirkwieser[1], Günther R. Raidl[1], and Jakob Puchinger[2]

[1] Institute of Computer Graphics and Algorithms
Vienna University of Technology, Vienna, Austria
{pirkwieser|raidl}@ads.tuwien.ac.at
[2] National ICT Australia (NICTA) Victoria Laboratory
Dep. of Comp. Sci. & Softw. Eng., The University of Melbourne, Australia
jakobp@csse.unimelb.edu.au

Abstract. We present a Lagrangian decomposition approach for the Knapsack Constrained Maximum Spanning Tree problem yielding upper bounds as well as heuristic solutions. This method is further combined with an evolutionary algorithm to a sequential hybrid approach. Experimental investigations, including a comparison to a previously suggested simpler Lagrangian relaxation based method, document the advantages of the new approach. Most of the upper bounds derived by Lagrangian decomposition are optimal, and together with the evolutionary algorithm, large instances with up to 12000 nodes can be either solved to provable optimality or with a very small remaining gap in reasonable time.

1 Introduction

The *Knapsack Constrained Maximum Spanning Tree* (KCMST) problem has been introduced by Yamamato and Kubo [1]. It arises in practice in certain situations where the aim is to design a profitable communication network under a strict limit on total costs for cable laying or similar resource constraints.

We are given an undirected connected graph $G = (V, E)$ with node set V and edge set $E \subseteq V \times V$ representing all possible connections. Each edge $e \in E$ has associated a weight $w_e \in \mathbb{Z}_+$ (corresponding to costs) and a profit $p_e \in \mathbb{Z}_+$. In addition, a weight limit (capacity) $c > 0$ is specified. A feasible KCMST is a spanning tree $T \subseteq E$ on G, i.e. a cycle-free subgraph connecting all nodes, whose weight $\sum_{e \in T} w_e$ does not exceed c. The objective is to find a KCMST with maximum total profit $\sum_{e \in T} p_e$. More formally, we can introduce binary variables x_e, $\forall e \in E$, indicating which edges are part of the solution, i.e. $x_e = 1 \leftrightarrow e \in T$ and $x_e = 0$ otherwise, and write the KCMST problem as:

$$\max \ p(x) = \sum_{e \in E} p_e x_e \tag{1}$$

C. Cotta and J. van Hemert (Eds.): EvoCOP 2007, LNCS 4446, pp. 176–187, 2007.

s. t. x represents a spanning tree $\qquad$ (2)

$$\sum_{e \in E} w_e x_e \leq c \qquad (3)$$

$$x_e \in \{0, 1\} \qquad\qquad \forall e \in E \qquad (4)$$

Obviously, the problem represents a combination of the classical minimum spanning tree problem (with changed sign in the objective function) and the classical 0–1 knapsack problem due to constraint (3). Yamada et al. [2] gave a proof for the KCMST problem's $\mathcal{NP}$-hardness.

After summarizing previous work for this problem in the next section, we present a Lagrangian decomposition approach in Section 3. It is able to yield tight upper bounds as well as lower bounds corresponding to feasible heuristic solutions. Section 4 describes an evolutionary algorithm for the KCMST problem utilizing the edge-set representation. Section 5 explains how this evolutionary algorithm can be effectively combined with the Lagrangian decomposition approach in a sequential manner. Experimental results are presented in Section 6. They document the excellent performance of the whole hybrid system, which is able to solve almost all test instances with graphs of up to 12000 nodes to provable optimality or with a very small gap in reasonable time.

2   Previous Work

While numerous algorithms and studies exist for the standard minimum spanning tree problem, the 0–1 knapsack problem, and various related constrained network design problems, we are only aware of the following literature specifically addressing the KCMST problem.

Yamamato and Kubo [1] introduced this problem, but neither proved $\mathcal{NP}$-hardness nor presented any solution algorithms. This was first done by Yamada et al. [2]. They described a Lagrangian relaxation approach in which the knapsack constraint (3) is relaxed, yielding the simple maximum spanning tree problem which can be solved efficiently. The Lagrangian dual problem of finding a best suited Lagrangian multiplier for the relaxed weight constraint is solved by a simple bisection method. The Lagrangian relaxation approach also yields feasible heuristic solutions, which are further improved by a 2-opt local search. In order to also determine provable optimal solutions for instances of restricted size, the Lagrangian relaxation is embedded in a branch-and-bound framework. While the approach is able to optimally solve instances with up to 1000 nodes and 2800 edges when edge weights and profits are uncorrelated, performance degrades substantially in the correlated case.

The only other work for the KCMST problem we are aware of is the first author's master thesis [3]. It formed the basis for this article, and we refer to it for further details, in particular for more computational results.

The problem also exists in its minimization version [4], for which Jörnsten and Migdalas document the superiority of Lagrangian decomposition, and subsequently solving each subproblem to optimality, for generating valid bounds [5].

3   Lagrangian Decomposition for the KCMST Problem

Lagrangian relaxation is a commonly used technique from the area of mathematical programming to determine upper bounds for maximization problems. Though the solutions obtained are in general infeasible for the original problem, they can lend themselves to create feasible solutions and thus to derive lower bounds, too. For a general introduction to Lagrangian relaxation, see [6,7,8]. *Lagrangian Decomposition* (LD) is a special variant that can be meaningful when there is evidence of two or possibly more intertwined subproblems, and each of them can be efficiently solved on its own by specialized algorithms.

As the KCMST problem is a natural combination of the maximum spanning tree problem and the 0–1 knapsack problem, we apply LD by aiming at such a partitioning. For this purpose, we split variables x_e, $\forall e \in E$, by introducing new variables y_e and including linking constraints, leading to the following equivalent reformulation:

$$\max \ p(x) = \sum_{e \in E} p_e x_e \tag{5}$$

$$\text{s. t. } \ x \text{ represents a spanning tree} \tag{6}$$

$$\sum_{e \in E} w_e y_e \leq c \tag{7}$$

$$x_e = y_e \qquad\qquad \forall e \in E \tag{8}$$

$$x_e, y_e \in \{0, 1\} \qquad\qquad \forall e \in E \tag{9}$$

The next step is to relax the linking constraints (8) in a Lagrangian fashion using Lagrangian multipliers $\lambda_e \in \mathbb{R}$, $\forall e \in E$. By doing so we obtain the Lagrangian decomposition of the original problem, denoted by KCMST-LD(λ):

$$\max \ p(x) = \sum_{e \in E} p_e x_e - \sum_{e \in E} \lambda_e(x_e - y_e) \tag{10}$$

$$\text{s. t. } \ x \text{ represents a spanning tree} \tag{11}$$

$$\sum_{e \in E} w_e y_e \leq c \tag{12}$$

$$x_e, y_e \in \{0, 1\} \qquad\qquad \forall e \in E \tag{13}$$

Stating KCMST-LD(λ) in a more compact way and emphasizing the now independent subproblems yields

$$(\text{MST}) \ \max \ \{(p - \lambda)^T x \mid x \ \hat{=} \ \text{a spanning tree}, x \in \{0,1\}^E\} \ + \tag{14}$$

$$(\text{KP}) \ \max \ \{\lambda^T y \mid w^T y \leq c, \ y \in \{0,1\}^E\}. \tag{15}$$

For a particular λ, the maximum spanning tree (MST) subproblem (14) can be efficiently solved by standard algorithms. In our implementation we apply

Kruskal's algorithm [9] based on a union-find data structure when the underlying graph is sparse and Prim's algorithm [10] utilizing a pairing heap with dynamic insertion [11] for dense graphs. The 0–1 knapsack subproblem (15) is known to be weakly $\mathcal{NP}$-hard, and practically highly efficient dynamic programming approaches exist [12], whereas we apply the COMBO algorithm [13].

It follows from Lagrangian relaxation theory that for any choice of Lagrangian multipliers λ, the optimal solution value to KCMST-LD(λ), denoted by v (KCMST- LD(λ)), is always at least as large as the optimal solution value of the original KCMST problem, i.e., KCMST-LD(λ) provides a valid upper bound. To obtain the tightest (smallest) upper bound, we have to solve the Lagrangian dual problem:

$$\min_{\lambda \in \mathbb{R}^E} v(\text{KCMST-LD}(\lambda)). \tag{16}$$

This dual problem is piecewise linear and convex, and standard algorithms like an iterative subgradient approach can be applied for (approximately) solving it. More specifically, we use the *volume algorithm* [14] which has been reported to outperform standard subgradient methods in many cases including set covering, set partitioning, max cut, and Steiner tree problems [15]. In fact, preliminary tests on the KCMST problem also indicated its superiority over a standard subgradient algorithm [3]. The volume algorithm's name is inspired by the fact that primal solutions are considered and that their values come from approximating the volumes below the active faces of the dual problem.

3.1   Strength of the Lagrangian Decomposition

According to integer linear programming theory, Lagrangian relaxation always yields a bound that is at least as good as the one obtained by the corresponding linear programming (LP) relaxation. The Lagrangian relaxation's bound can be substantially better when the relaxed problem does not fulfill the integrality property, i.e., the solution to the LP relaxation of the relaxed problem – KCMST-LD(λ) in our case – is in general not integer.

For seeing whether or not this condition is fulfilled here, we have to consider both independent subproblems. Compact models having the integrality property exist for MST, see e.g. [16]. Furthermore, the integrality property is obviously not fulfilled for the knapsack subproblem. Thus, we may expect to obtain bounds that are better than those from the linear programming relaxation of KCMST.

In comparison, in the Lagrangian relaxation approach from [2] the knapsack constraint is relaxed and only the MST problem remains. This approach therefore fulfills the integrality property and, thus, is in general weaker than our LD.

We further remark that the proposed LD can in principle be strengthened by adding the cardinality constraint $\sum_{e \in E} y_e = |V| - 1$ to the knapsack subproblem. The resulting cardinality constrained knapsack problem is still only weakly $\mathcal{NP}$-hard, and pseudo-polynomial algorithms based on dynamic programming are known for it [12]. Our investigations indicate, however, that the computational demand required for solving this refined formulation is in practice substantially higher and does not pay off the typically only small quality increase of the obtained bound [3].

3.2   Deriving Lower Bounds

In some iterations of the volume algorithm, the obtained spanning tree is feasible with respect to the knapsack constraint and can be directly used as a lower bound, hence resulting in a simple Lagrangian heuristic. In order to further improve such solutions this heuristic is strengthened by consecutively applying a local search based on the following edge exchange move.

1. Select an edge $(u, v) \in E \setminus T$ to be considered for inclusion (see below).
2. Determine the path $P \subseteq T$ connecting nodes u and v in the current tree. Including e in T would yield the cycle $P \cup \{(u, v)\}$.
3. Identify a least profitable edge $\tilde{e} \in P$ that may be replaced by (u, v) without violating the knapsack constraint:

$$\tilde{e} = \text{minarg} \ \{p_e \mid e \in E \wedge w(T) - w_e + w_{(u,v)} \leq c\}, \qquad (17)$$

where $w(T) = \sum_{e \in T} w_e$. In case of ties, an edge with largest weight is chosen.
4. If replacing $\tilde{e}$ by (u, v) improves the solution, i.e. $p_{\tilde{e}} < p_{(u,v)} \vee (p_{\tilde{e}} = p_{(u,v)} \wedge w_{\tilde{e}} > w_{(u,v)})$, perform this exchange.

For selecting edge (u, v) in step 1 we consider two possibilities:

Random selection: Randomly select an edge from $E \setminus T$.
Greedy selection: At the beginning of the local search, all edges are sorted according to decreasing $p'_e = p_e - \lambda_e$, the reduced profits used to solve the MST subproblem. Then, in every iteration of local search, the next less profitable edge not active in the current solution is selected. This results in a greedy search where every edge is considered at most once.

Since Lagrangian multipliers are supposed to be of better quality in later phases of the optimization process, local search is only applied when the ratio of the incumbent lower and upper bounds is larger than a certain threshold τ. Local search stops after ρ consecutive non-improving iterations have been performed.

4   A Suitable Evolutionary Algorithm

Evolutionary algorithms (EAs) have often proven to be well suited for finding good approximate solutions to hard network design problems. In particular for constrained spanning tree problems, a large variety of EAs applying very different representations and variation operators have been described, see e.g. [17] for an overview.

Here, we apply an EA based on a direct edge-set representation for heuristically solving the KCMST problem, since this encoding and its corresponding variation operators are known to provide strong locality and heritability. Furthermore, variation operators can efficiently be applied in time that depends (almost) only linearly on the number of nodes. In fact, our EA closely follows the description of the EA for the degree constrained minimum spanning tree problem in [17]. Only the initialization and variation operators are adapted to conform with the knapsack constraint.

The general framework is steady-state, i.e. in each iteration one feasible off-spring solution is created by means of recombination, mutation, and eventually local improvement, and it replaces the worst solution in the population. Duplicates are not allowed in the population; they are always immediately discarded. The EA's operators work as follows.

Initialization. To obtain a diversified initial population, a random spanning tree construction based on Kruskal's algorithm is used. Edges are selected with a bias towards those with high profits. The specifically applied technique is exactly as described in [17]. In case a generated solution is infeasible with respect to the knapsack constraint, it is stochastically repaired by iteratively selecting a not yet included edge at random, adding it to the tree, and removing an edge with highest weight from the induced cycle.

Recombination. An offspring is derived from two selected parental solutions in such a way that the new solution candidate always exclusively consists of inherited edges: In a first step all edges contained in both parents are immediately adopted. The remaining parental edges are merged into a single candidate list. From this list, we iteratively select edges by binary tournaments with replacement favoring high-profit edges. Selected edges are included in the solution if they do not introduce a cycle; otherwise, they are discarded. The process is repeated until a complete spanning tree is obtained. Finally, its validity with respect to the knapsack constraint is checked. An infeasible solution is repaired in the same way as during initialization, but only considering parental edges for inclusion.

Mutation. We perform mutation by inserting a randomly selected new edge and removing another edge from the introduced cycle. The choice of the edge to be included is biased towards high-profit edges by utilizing a normally-distributed rank-based selection as described in [17]. The edge to be removed from the induced cycle is chosen at random among those edges whose removal would retain a feasible solution.

Local Search. With a certain probability, a newly derived candidate solution is further improved by the local search procedure described in Section 3.2.

5   Hybrid Lagrangian Evolutionary Algorithm

Preliminary tests clearly indicated that the EA cannot compete with the performance of LD in terms of running time and solution quality. However, following similar ideas as described in [15] for the price-collecting Steiner tree problem, we can successfully apply the EA for finding better final solutions after performing LD. Hereby, the EA is adapted to exploit a variety of (intermediate) results from LD. In detail, the following steps are performed after LD has terminated and before the EA is executed:

1. If the profit of the best feasible solution obtained by LD corresponds to the determined upper bound, we already have an optimal solution. No further actions are required.

2. For the selection of edges during initialization, recombination, and mutation of the EA, original edge profits p_e are replaced by reduced profits $p'_e = p_e - \lambda_e$. In this way, Lagrangian dual variables are exploited, and the heuristic search emphasizes the inclusion of edges that turned out to be beneficial in LD.

3. The edge set to be considered by the EA is reduced from E to a subset E' containing only those edges that appeared in any of the feasible solutions encountered by LD. For this purpose, LD is extended to mark these edges.

4. The best feasible solution obtained by LD is included in the EA's initial population.

5. Finally, the upper bound obtained by LD is passed to the EA and exploited by it as an additional stopping criterion: When a solution with a corresponding total profit is found, it is optimal and the EA terminates.

6   Experimental Results

The described algorithms have been tested on a large variety of different problem instances, and comparisons have been performed in particular with the previous Lagrangian relaxation based method from [2]. This section summarizes most important results; more details can be found in [3]. All experiments were run on a 1.6GHz Pentium M PC with 1.25GB RAM.

As in [2], we consider instances based on random complete graphs $K_{|V|\gamma}$ and planar graphs $P_{|V|,|E|\gamma}$. Since we could not obtain the original instances, we created them in the same way by our own. In addition we constructed larger maximal planar graphs $P_{|V|\gamma}$. Parameter γ represents the type of correlation between profits and weights:

uncorrelated ('u'): p_e and w_e, $e \in E$, are independently chosen from the integer interval $[1, 100]$;

weakly correlated ('w'): w_e is chosen as before, and $p_e := \lfloor 0.8w_e + v_e \rfloor$, where v_e is randomly selected from $[1, 20]$;

strongly correlated ('s'): w_e is chosen as before, and $p_e := \lfloor 0.9w_e + 10 \rfloor$.

For details on the methods used to construct the (maximal) planar graphs, we refer to [2,3]. In case of complete graphs, the knapsack capacity is $c = 20 \cdot |V| - 20$, in case of (maximal) planar graphs $c = 35 \cdot |V|$. For each combination of graph type, graph size, and correlation, 10 instances have been considered.

We show and compare results for the Lagrangian relaxation (LR), Lagrangian relaxation with local search (LR+LS), and associated branch-and-bound (B&B) from [2], our Lagrangian decomposition with the simple primal heuristic (LD) and optionally local search (LD+LS), and the combination of LD and the EA (LD+LS+EA).

Robust settings for strategy parameters have been determined by preliminary tests. For the results presented here the following setup has been used.

The volume algorithm within the LD approach terminates when either the lower and upper bounds become identical and, thus, an optimal solution has been reached, or when the upper bound did not improve over the last 500 iterations in case of planar graphs and 1000 iterations in case of complete graphs. For completeness, we provide the following further details for the volume algorithm based on its description in [14]: The target value T always is updated by $T :=$ $0.95LB$ and $T := 0.475(LB+UB)$ for planar and complete graphs, respectively, with the exception $T := 0.95T$ iff $UB < 1.05T$. Parameter f is initialized with 0.1 and multiplied by 0.67 after 20 consecutive *red* iterations when $f > 10^{-8}$ in case of planar graph and $f > 10^{-6}$ for complete graphs and is multiplied by 1.1 after a *green* iteration when $f < 1$. Factor α is initialized with 0.1 and it is checked after every 100 and 200 iterations in case of planar and complete graphs, respectively, if the upper bound decreased less than 1%; if so and $\alpha > 10^{-5}$ then $\alpha := 0.85\alpha$. All these update rules are similar to those used in [15].

For the optional local search, greedy edge selection is used for complete graphs and random edge selection for all others. The application threshold is set to $\tau = 0.99$. As maximum number of iterations without improvement, $\rho = 200$ is used in case of uncorrelated and weakly correlated planar graphs, and $\rho = 100$ in all other cases.

For the EA, the population size is 100, binary tournament selection is used, and recombination and mutation are always applied. For the biasing towards edges with higher profits, parameters α and β (see [17]) are both set to 1.5. Local search is performed with random edge selection for each new candidate solution with a probability of 20% with $\rho = 50$ and a maximum of 5000 total iterations for graphs having less than 8000 nodes and 10000 total iterations for larger graphs.

Results on planar and complete graphs are shown in Table 1. For LR, LR+LS, and B&B, they are adopted from [2]. Average values based on 10 different instances are printed. Columns LB show obtained lower bounds, i.e. the objective values of the best feasible solutions. Upper bounds (UB) are expressed in terms of the relative gap to these lower bounds: $gap = (UB - LB)/LB$; corresponding standard deviations are listed in columns σ_{gap}. Columns Opt show numbers of instances (out of 10) for which the gap is zero and, thus, optimality has been proven. Average CPU-times for the runs are printed in columns t in seconds, and the average numbers of iterations of the volume algorithm in columns $iter$.

With respect to the CPU-times listed for branch-and-bound, we remark that they were measured on an IBM RS/6000 44P Model 270 workstation, and therefore, they cannot directly be compared with the times from our methods. The maximum time limit for B&B was 2000 seconds.

Most importantly, we can see that LD obtains substantially smaller gaps than both, LR and LR+LS. In fact, LD's average gaps are never larger than 0.063%, and for a large number of instances, optimality is already proven. On the remaining instances, enhancing LD by applying local search is beneficial; in most cases gaps are significantly reduced, and a few more instances could be solved

Table 1. Results of Lagrangian algorithms on planar and complete graphs

Instance	Yamada et al. [2]				LD						LD+LS					
	LR gap $[\cdot 10^{-5}]$	LR+LS gap $[\cdot 10^{-5}]$	B&B $t[s]$	Opt	$t[s]$	iter	LB	gap $[\cdot 10^{-5}]$	σ_{gap} $[\cdot 10^{-5}]$	Opt	$t[s]$	iter	LB	gap $[\cdot 10^{-5}]$	σ_{gap} $[\cdot 10^{-5}]$	Opt
$P_{50,127u}$	948.2	454.1	0.43	10	0.19	983	3558.5	62.56	89.70	3	0.30	976	3559.0	**47.58**	49.16	3
$P_{100,260u}$	586.6	268.9	1.78	10	0.17	801	7222.9	**6.76**	13.17	7	0.37	817	7222.9	**6.76**	13.17	7
$P_{200,560u}$	411.6	187.9	5.46	10	0.31	869	14896.7	3.98	5.60	6	0.55	822	14896.9	**2.68**	4.71	7
$P_{400,1120u}$	128.3	70.4	24.44	10	0.55	880	29735.0	2.71	3.83	6	1.15	905	29735.1	**2.36**	3.20	6
$P_{600,1680u}$	121.2	54.1	75.25	10	0.79	934	44836.2	1.11	1.17	5	1.52	854	44836.4	**0.67**	1.07	7
$P_{800,2240u}$	296.2	124.9	466.37	10	0.79	766	59814.5	0	0	10	1.59	716	59814.5	0	0	10
$P_{1000,2800u}$	166.0	73.3	592.77	10	0.99	764	74835.6	0	0	10	2.08	764	74835.6	0	0	10
$P_{50,127w}$	4372.0	1243.3	0.81	10	0.15	745	2063.2	52.80	79.75	6	0.23	751	2063.6	**33.57**	50.59	6
$P_{100,260w}$	2926.4	603.7	2.71	10	0.17	732	4167.9	9.67	16.94	7	0.36	724	4168.0	**7.24**	11.65	7
$P_{200,560w}$	1064.0	266.3	13.11	10	0.28	730	8431.9	1.19	3.76	9	0.36	634	8432.0	0	0	10
$P_{400,1120w}$	818.8	183.9	47.15	10	0.49	802	16794.3	3.58	6.42	7	0.77	721	16794.9	0	0	10
$P_{600,1680w}$	824.0	167.6	371.84	8	0.65	779	25158.0	**0.40**	1.26	9	1.29	788	25158.0	**0.40**	1.26	9
$P_{800,2240w}$	425.7	103.8	509.22	5	0.92	854	33540.2	0.89	1.99	8	1.76	762	33540.5	0	0	10
$P_{50,127s}$	10282.5	161.0	2.84	10	0.16	815	2051.3	43.92	62.81	5	0.12	573	2052.2	0	0	10
$P_{100,260s}$	19898.0	265.6	405.45	8	0.23	829	4115.1	9.72	12.54	6	0.18	641	4115.5	0	0	10
K_{40u}	250.9	106.1	0.87	10	0.23	880	3669.3	**5.50**	11.59	8	0.28	884	3669.3	**5.50**	11.59	8
K_{60u}	390.1	107.4	1.89	10	0.58	1164	5673.3	8.86	12.50	6	0.72	1189	5673.4	**7.10**	9.16	6
K_{80u}	272.7	130.3	6.54	10	0.60	858	7672.8	0	0	10	0.69	847	7672.8	0	0	10
K_{100u}	148.8	43.3	12.48	10	1.07	1062	9698.0	**1.03**	3.25	9	1.27	1055	9698.0	**1.03**	3.25	9
K_{120u}	122.3	42.7	23.69	10	1.37	1012	11701.2	0	0	10	1.65	1052	11701.2	0	0	10
K_{140u}	56.1	22.6	60.95	10	2.08	1184	13721.0	0	0	10	2.38	1162	13721.0	0	0	10
K_{160u}	89.7	38.8	476.26	10	2.88	1260	15727.9	0	0	10	3.19	1213	15727.9	0	0	10
K_{180u}	101.1	45.2	636.54	10	4.31	1488	17729.2	1.13	3.57	9	4.95	1470	17729.3	0.56	1.77	9
K_{200u}	40.5	17.2	375.26	10	5.55	1502	19739.4	0	0	10	6.11	1446	19739.4	0	0	10
K_{20w}	6186.9	991.7	0.25	10	0.11	720	618.9	**17.01**	53.79	9	0.12	698	618.9	**17.01**	53.79	9
K_{40w}	4262.5	520.3	1.17	10	0.24	737	1320.6	7.55	23.87	9	0.19	613	1320.7	0	0	10
K_{60w}	5700.5	529.2	6.09	10	0.51	891	2017.6	19.87	41.88	8	0.40	676	2018.0	0	0	10
K_{80w}	4970.4	343.6	38.15	10	0.81	863	2720.4	3.68	11.63	9	0.67	732	2720.5	0	0	10
K_{100w}	2413.3	172.9	377.61	8	1.10	879	3421.3	2.92	9.23	9	1.02	759	3421.4	0	0	10
K_{120w}	3797.7	206.6	451.06	8	2.78	1527	4123.3	26.69	24.15	3	1.65	871	4124.3	2.43	7.68	9
K_{20s}	22122.2	379.1	0.53	10	0.22	960	528.6	56.89	91.60	7	0.09	635	528.9	0	0	10
K_{30s}	17032.9	322.2	99.63	10	0.31	1016	809.2	37.12	59.76	7	0.16	717	809.5	0	0	10
K_{40s}	9492.7	137.7	226.30	6	0.34	902	1089.9	18.38	58.12	9	0.28	782	1090.1	0	0	10

to proven optimality. Overall, only 40 out of 330 instances remain, for which LD+LS was not able to find optimal solutions and prove their optimality. As already observed in [2], strongly correlated instances are typically harder to solve than uncorrelated ones.

A comparison of the heuristic solutions obtained from LD+LS with solutions from an exact approach[1] further indicated that almost all of them are actually optimal; LD+LS just cannot prove their optimality since the upper bounds were not tight enough. As a consequence, additionally applying the EA after LD+LS was not very meaningful for these instances. Tests not shown here confirmed that only in rare cases, gaps could further be reduced by the EA.

Our LD is extremely fast, needing for none of these instances more than seven seconds. The time overhead introduced by local search is also only very moderate, in particular since the improved heuristic solutions implied a faster convergence of the volume algorithm.

---

[1] We also implemented a not yet published exact branch-and-cut algorithm, which is able to solve these instances to proven optimality.

Table 2. Results of Lagrangian and hybrid algorithms on maximal planar graphs

Instance	LD						LD+LS						LD+LS+EA								
	t[s]	iter	LB	gap [·10⁻⁵]	σ_gap [·10⁻⁵]	Opt	t[s]	iter	LB	gap [·10⁻⁵]	σ_gap [·10⁻⁵]	Opt	t[s]	iter	red	iter_EA	LB	gap [·10⁻⁵]	σ_gap [·10⁻⁵]	Opt	Opt_EA
P_{2000u}	2.32	867	147799.4	0.14	0.29	8	3.26	813	147799.6	**0**	0	10	4.34	816	38%	2188	147799.6	**0**	0	10	1
P_{2000w}	2.42	883	85570.1	0.81	1.09	6	3.29	808	85570.7	0.12	0.37	9	6.29	856	44%	2001	85570.8	**0**	0	10	3
P_{2000s}	2.97	1045	82520.9	2.90	3.09	2	2.87	815	82523.3	**0**	0	10	3.33	816	20%	0	82523.3	**0**	0	10	0
P_{4000u}	4.64	854	294872.0	0.03	0.09	9	7.37	835	294872.0	**0.03**	0.09	9	12.42	853	39%	5000	294872.0	**0.03**	0.09	9	0
P_{4000w}	5.44	1040	170957.1	0.60	0.48	3	7.77	907	170957.7	0.24	0.50	8	12.93	985	43%	1283	170958.1	**0**	0	10	3
P_{4000s}	6.10	1071	165048.9	1.57	1.63	2	7.92	916	165051.4	0.06	0.18	9	7.53	887	23%	0	165051.5	**0**	0	10	0
P_{6000u}	8.16	953	441977.5	0.13	0.22	6	12.46	898	441978.1	**0**	0	10	23.45	959	39%	2674	441978.1	**0**	0	10	2
P_{6000w}	9.12	1033	256316.7	0.67	0.84	4	12.66	934	256318.3	0.04	0.12	9	20.93	980	45%	1130	256318.4	**0**	0	10	3
P_{6000s}	9.94	1094	247588.6	1.45	1.87	2	12.68	950	247592.2	**0**	0	10	15.62	937	25%	1325	247592.2	**0**	0	10	1
P_{8000u}	11.15	906	589446.9	0.04	0.08	8	19.58	975	589446.9	0.04	0.08	8	17.88	892	39%	0	589447.1	**0**	0	10	0
P_{8000w}	13.89	1102	341901.7	0.80	0.86	3	19.49	981	341904.0	0.12	0.20	7	26.97	919	46%	3503	341904.4	**0**	0	10	1
P_{8000s}	14.22	1087	330117.3	1.44	1.45	3	17.02	887	330122.0	**0.03**	0.09	9	39.37	922	23%	3968	330122.0	**0.03**	0.09	9	1
P_{10000u}	15.92	969	737450.2	0.07	0.12	7	24.63	956	737450.6	0.01	0.03	9	56.66	1029	39%	1877	737450.7	**0**	0	10	4
P_{10000w}	16.31	964	427406.4	0.19	0.39	7	25.51	1021	427406.9	0.06	0.09	7	61.62	1048	44%	1681	427407.2	**0**	0	10	5
P_{10000s}	23.42	1383	412640.1	0.84	0.87	1	26.61	1025	412643.6	**0**	0	10	26.82	1019	23%	0	412643.6	**0**	0	10	0
P_{12000u}	21.67	1056	885117.0	0.08	0.10	5	29.20	921	885117.8	**0**	0	10	55.54	1008	39%	1468	885117.8	**0**	0	10	3
P_{12000w}	23.27	1102	512985.4	0.38	0.48	3	32.69	1033	512986.9	0.08	0.13	7	77.05	1037	45%	2147	512987.3	**0**	0	10	4
P_{12000s}	25.83	1148	495164.0	1.14	1.38	2	34.38	1019	495169.5	0.04	0.08	8	141.99	1044	23%	8225	495169.6	**0.02**	0.06	9	1

In order to investigate the usefulness of the proposed LD+LS+EA hybrid, we now turn to the larger maximal planar graphs, for which Table 2 presents results. For the EA, we additionally list the average number of EA iterations $iter_{EA}$, the relative amount of edges discarded after performing LD $red = (|E| - |E'|)/|E| \cdot 100\%$, and the number of optimal solutions Opt_{EA}, among Opt, found by the EA.

Again, the solutions obtained by LD are already quite good and gaps are in general small. The inclusion of local search clearly increases the number of optimal solutions found, leaving only 21 out of all 180 instances for which optimality is not yet proven. The hybrid approach (LD+LS+EA) works almost perfectly: Gaps are reduced to zero, and thus proven optimal solutions are achieved for all but three instances. The values in column Opt_{EA} document that the EA plays a significant role in finally closing gaps. The three remaining instances are solved with gaps less than 0.00003%.

In general, results of Tables 1 and 2 indicate that it is harder to close the optimality gap for smaller than for larger instances. One reason seems to be that with increasing graph size, more edges have the same profit and weight values. Tests on other types of instances, with differently determined profits and weights, are therefore interesting future work.

7 Conclusions

We presented a Lagrangian decomposition approach for the $\mathcal{NP}$-hard KCMST problem to derive upper bounds as well as heuristic solutions. Experimental results on large graphs revealed that the upper bounds are extremely tight, in fact most of the time even optimal. Heuristic solutions can be significantly improved by applying a local search, and many instances can be solved to provable optimality already in this way.

For the remaining, larger instances, a sequential combination of LD with an evolutionary algorithm has been described. The EA makes use of the edge-set encoding and corresponding problem-specific operators and exploits results from LD in several ways. In particular, the graph is shrunk by only considering edges also appearing in heuristic solutions of LD, Lagrangian dual variables are exploited by using final reduced costs for biasing the selection of edges in the EA's operators, and the best solution obtained from LD is provided to the EA as seed in the initial population.

Computational results document the effectiveness of the hybrid approach. The EA is able to close the gap and provide proven optimal solutions in almost all of the remaining difficult cases. Hereby, the increase in running time one has to pay is only moderate.

The logical next step we want to pursue is to enhance the branch-and-bound method from [2] by also utilizing the more effective LD or even the hybrid LD/EA instead of the simple Lagrangian relaxation.

In general, we believe that such combinations of Lagrangian relaxation and metaheuristics like evolutionary algorithms are highly promising for many combinatorial optimization tasks. Future work therefore includes the consideration of further problems, but also the closer investigation of other forms of collaboration between Lagrangian relaxation based methods and metaheuristics, including intertwined and parallel models.

References

1. Yamamato, Y., Kubo, M.: Invitation to the Traveling Salesman's Problem (in Japanese). Asakura, Tokyo (1997)
2. Yamada, T., Watanabe, K., Katakoa, S.: Algorithms to solve the knapsack constrained maximum spanning tree problem. Int. Journal of Computer Mathematics **82**(1) (2005) 23–34
3. Pirkwieser, S.: A Lagrangian Decomposition Approach Combined with Metaheuristics for the Knapsack Constrained Maximum Spanning Tree Problem. Master's thesis, Vienna University of Technology, Institute of Computer Graphics and Algorithms (October 2006)
4. Aggarwal, V., Aneja, Y., Nair, K.: Minimal spanning tree subject to a side constraint. Comput. & Operations Res. **9**(4) (1982) 287–296
5. Jörnsten, K., Migdalas, S.: Designing a minimal spanning tree network subject to a budget constraint. Optimization **19**(4) (1988) 475–484
6. Fisher, M.L.: The Lagrangian Relaxation Method for Solving Integer Programming Problems. Management Science **27**(1) (1981) 1–18
7. Fisher, M.L.: An application oriented guide to Lagrangean Relaxation. Interfaces **15** (1985) 10–21
8. Beasley, J.E.: Lagrangian relaxation. In Reeves, C.R., ed.: Modern Heuristic Techniques for Combinatorial Problems. John Wiley & Sons, Inc., New York (1993) 243–303
9. Kruskal, J.B.: On the shortest spanning subtree of a graph and the travelling salesman problem. In: Proc. of the AMS. Volume 7. (1956) 48–50
10. Prim, R.C.: Shortest connection networks and some generalizations. Bell Systems Technology Journal **36** (1957) 1389–1401
11. Fredman, M.L., Sedgewick, R., Sleator, D.D., Tarjan, R.E.: The pairing heap: a new form of self-adjusting heap. Algorithmica **1**(1) (1986) 111–129
12. Kellerer, H., Pferschy, U., Pisinger, D.: Knapsack Problems. Springer Verlag (2004)
13. Martello, S., Pisinger, D., Toth, P.: Dynamic programming and strong bounds for the 0–1 knapsack problem. Management Science **45** (1999) 414–424
14. Barahona, F., Anbil, R.: The volume algorithm: producing primal solutions with a subgradient method. Mathematical Programming **87**(3) (2000) 385–399
15. Haouaria, M., Siala, J.C.: A hybrid Lagrangian genetic algorithm for the prize collecting Steiner tree problem. Comput. & Operations Res. **33**(5) (2006) 1274–1288
16. Magnanti, T.L., Wolsey, L.A.: Optimal trees. In Ball, M.O., et al., eds.: Handbooks in Operations Research and Management Science. Volume 7. Elsevier Science (1995) 503–615
17. Julstrom, B.A., Raidl, G.R.: Edge sets: an effective evolutionary coding of spanning trees. IEEE Transactions on Evolutionary Computation **7**(3) (2003) 225–239

Exact/Heuristic Hybrids Using rVNS and Hyperheuristics for Workforce Scheduling*

Stephen Remde, Peter Cowling, Keshav Dahal, and Nic Colledge

MOSAIC Research Group, University of Bradford, Great Horton Road
Bradford, BD7 1DP, United Kingdom
{s.m.remde, p.i.cowling, k.p.dahal, n.j.colledge}@bradford.ac.uk

Abstract. In this paper we study a complex real-world workforce scheduling problem. We propose a method of splitting the problem into smaller parts and solving each part using exhaustive search. These smaller parts comprise a combination of choosing a method to select a task to be scheduled and a method to allocate resources, including time, to the selected task. We use reduced Variable Neighbourhood Search (rVNS) and hyperheuristic approaches to decide which sub problems to tackle. The resulting methods are compared to local search and Genetic Algorithm approaches. Parallelisation is used to perform nearly one CPU-year of experiments. The results show that the new methods can produce results fitter than the Genetic Algorithm in less time and that they are far superior to any of their component techniques. The method used to split up the problem is generalisable and could be applied to a wide range of optimisation problems.

1 Introduction

In collaboration with an industrial partner we have studied a workforce scheduling problem which is a resource constrained scheduling problem similar to but more complex than many other well-studied scheduling problems such as the Resource Constrained Project Scheduling Problem (RCPSP) [1] and job shop scheduling problem [2]. The problem is based on our work with @Road Ltd. which develops scheduling solutions for very large, complex mobile workforce scheduling problems in a variety of industries. Our workforce scheduling problem is concerned with assigning people and other resources to geographically dispersed tasks while respecting time window constraints and skill requirements.

The workforce scheduling problem that we consider consists of four main components: Tasks, Resources, Skills and Locations. Unlike many RCPSP problems, the tasks have locations and a priority value (to indicate relative importance). Resources are engineers and large pieces of equipment. They are mobile, travelling at a variety of speeds to geographically dispersed tasks. Tasks and resources have time windows with different associated costs (to consider, for example, inconvenience to

* This work was funded by EPSRC and @Road Ltd under an EPSRC CASE studentship, which was made available through and facilitated by the Smith Institute for Industrial Mathematics and System Engineering.

C. Cotta and J. van Hemert (Eds.): EvoCOP 2007, LNCS 4446, pp. 188–197, 2007.

customers at certain times, the cost of overtime, etc.). Tasks require a specified amount of specified skills, and resources possess one or more of these skills at different competencies which affects the amount of time required. A major source of complexity of our problem comes from the fact that a task's duration is unknown until resources are assigned to it. In this paper, the fitness of a schedule is given by one of the single weighted objective functions used in [3], $f = SP - 4SC - 2TT$, where SP is the sum of the priority of scheduled tasks, SC is the sum of the time window costs in the schedule (both resource and task) and TT is the total amount of travel time. This objective is to maximise the total priority of tasks scheduled while minimising travel time and cost. [3] describes the problem in more detail and uses a Genetic Algorithm to solve it. In this paper we will compare the Genetic Algorithm method with a new reduced Variable Neighbourhood Search and hyperheuristic methods.

We propose a method to break down this "messy" problem by splitting it into smaller parts and solving each part using exact enumerative approaches. Hence each part consists of finding the optimal member of a local search neighbourhood. We then design ways to decide which part to tackle at each stage in the solution process. These smaller parts are the combination of a method to select a task and a method to select resources for the task. We will take these smaller parts and use reduced Variable Neighbourhood Search and hyperheuristics to decide the order in which to solve them.

This paper is structured as follows: we present related work in section 2 and propose reduced Variable Neighbourhood Search and hyperheuristic approaches in section 3. In section 4 we empirically investigate the new techniques and compare them to a genetic algorithm in terms of solution quality and computational time. We present conclusions in section 5.

2 Related Work

The RCPSP [1] involves a set of tasks which have to be scheduled under resource and precedence constraints. Precedence constraints require that a task may not start until all its preceding tasks have finished. Resource constraints require specified amounts of finite resources to be available when the task is scheduled. Scheduling an RCPSP involves assigning start times to each of the tasks. The RCPSP is a generalisation of many scheduling problems including job-shop, open-shop and flow-shop scheduling problems. The RCPSP has no notion of variable time dependant on skill or location of tasks and resources. The time line is also discrete and assumes resources are always available.

The Multimode Resource Constrained Resource Scheduling Problem (MRCPSP) extends the RCPSP [4]. In the MRCPSP, there is the option of having non-renewable resources and resources that are only available during certain periods. In addition, a task maybe executed in one of several execution modes. Each execution mode has different resource requirements and different task durations. Usually the number of these modes is small and hence exact methods can be used. In the workforce scheduling problem considered in this paper, we have a very large number of execution modes (as the task duration depends on the resource competency and the task skill requirement which are both real values). [5] uses a genetic algorithm as a solution to problems where using an exact method is intractable. [6] surveys heuristic solutions to the RCPSP and MRCPSP.

Solution methods such as Genetic Algorithms (GAs) were introduced by Bremermann [7] and the seminal work done by Holland [8]. Since then they have been developed extensively to tackle problems including the travelling salesman problem [9], bin packing problems [10] and scheduling problems [11]. A Genetic Algorithm tries to evolve a population into fitter ones by a process analogous to evolution in nature. Our previous work [3] compares a multi-objective genetic algorithm to a single weight objective genetic algorithm to study the trade-off between diversity and solution quality. The genetic algorithm is used to solve the dynamic workforce scheduling problem studied in this paper.

Variable Neighbourhood search (VNS) is a relatively new search technique and the seminal work was done by Mladenović and Hansen [12]. VNS is based on the idea of systematically changing the neighbourhood of a local search algorithm. Variable Neighbourhood Search enhances local search using a variety of neighbourhoods to "shake" the search into a new position after it reaches a local optimum. Several variants of VNS exist as extensions to the VNS framework [13].

Reduced Variable Neighbourhood search (rVNS) [13] is an attempt to improve the speed of variable neighbourhood search (with the possibility of a worse solution). Usually, the most time consuming part of VNS is the local search. rVNS picks solutions randomly from neighbourhoods which provide progressively larger moves. rVNS is targeted at large problems where computational time is more important than the quality of the result. In combinatorial optimisation problems, local search moves like "swap two elements" are frequently used, and [14] for RCPSP as well as others such as [15], apply VNS by having the neighbourhoods make an increasing number of consecutive local search moves. [16] however defines only two neighbourhoods for VNS applied to the Job Shop Scheduling Problem, a swap move and an insert move, which proves to be effective.

VNS can be seen as a form of hyperheuristic where the neighbourhoods and local search are low level heuristics. The term "hyperheuristic" was introduced in [17]. Hyperheuristics rely on low level heuristics and objective measures which are specific to the problem. The hyperheuristic uses feedback from the low level heuristics (CPU time taken, change in fitness, etc.) and determines which low level heuristics to use at each decision point. Earlier examples of hyperheuristics include [18] where a genetic algorithm evolves a chromosome which determined how jobs were scheduled in open shop scheduling. A variety of hyperheuristics have been developed including a learning approach based on the "choice function" [17], tabu search [19], simulated annealing [20] and Genetic Algorithms [21].

3 Heuristic Approaches

Our proposed framework splits the problem into (1) selecting a task to be scheduled and (2) selecting potential resources for that task. A task is randomly chosen from the top two tasks which we have not tried to schedule ranked by the task order, to make the search stochastic, to ensure that running it multiple times will produce different results. We have implemented 8 task selection methods given in table 1. Note that some of our task orders are deliberately counterintuitive to give us a basis for comparison.

Table 1. Task sorting methods

Method	Description
Random	Tasks are ordered at random.
PriorityDesc	Tasks are ordered by their priority in descending order
PriorityAsc	Tasks are ordered by their priority in ascending order
PrecedenceAsc	Tasks are ordered by their number of precedences ascending
PrecedenceDesc	Tasks are ordered by their number of precedences descending
PriOverReq	Tasks are ordered by their estimated priority per hour assuming the task will take as long as the total skill requirement
PriOverMaxReq	Tasks are ordered by their estimated priority per hour assuming the task will take as long as the maximum skill requirement
PriOverAvgReq	Tasks are ordered by their estimated priority per hour assuming the task will take as long as the average skill requirement

PriorityDesc, PriOverReq, PriOverMaxReq and *PriOverAvgReq* are attempts to identify the tasks which will give us the most reward and schedule them first. They estimate the task duration differently and use this estimate to calculate priority hour. *PrecedenceDesc* attempts to schedule those tasks with the largest number of succeeding tasks first. *PrecedenceAsc, PriorityAsc* and *Random* give us some indication of the effect of task orders since intuition would suggest that they should give poor results.

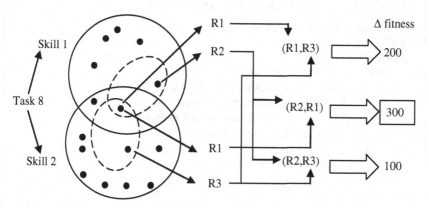

Fig. 1. Resource Selector. The dotted subset of resources possessing the required skill is chosen by a Resource Selector. The assignment (R2, R1) is chosen as the best insertion.

We then define Resource Selectors which select a set of potential resources for each skill required by the selected task. The Resource Selectors first sort the resources by their competencies at the skill required and then select a subset of them. This could be, for example, the top five or the top six to ten etc. The subsets of resources are then enumerated and exhaustive search used to find the insertion which will yield the lowest time window and travel penalties subject to precedence constraints. Figure 1 illustrates this.

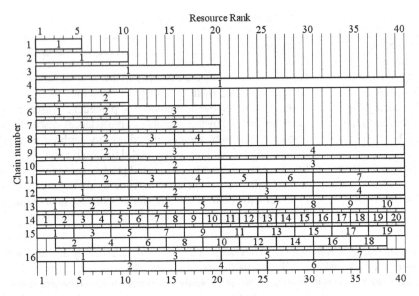

Fig. 2. Resource selection "chains" for the rVNS

k is the index of the resource selector in use
$(N_1, N_2, \ldots N_{k_{max}})$ is our chain of resource selectors

Sort tasks using the chosen task order
$k:=1$
while ($k < k_{max}$)
    for each Unscheduled Task T
        Select Sets of Resources Using N_k for Task T
        Exhaustively Search the selected sets of resources
                to find an optimal insertion I
        Insert task T into the schedule using I
    next
    if *some tasks were inserted* then
        $k:=1$
    else
        $k:=k+1$
    end if
end while

Fig. 3. Pseudo code for our rVNS method

The neighborhoods of our rVNS insert tasks selected by a task order using a given resource selector. If an insertion is not possible, because of resource or task constraints, we try the next resource selector and so on. We consider several sequences of resource selection neighborhoods, or "chains", as shown in figure 2. These neighborhoods show a progression of an increasing range of resources used and smaller sets. Figure 3 shows the pseudo code for our rVNS method. Allowing search to restart at the start of the chain allows the search to retry insertions that may have failed before because of resource or task constraints. If s is the maximum number of skills and n is the number of tasks, then the algorithm has time complexity $O(nk|N_k|^s)$ for each chain. With the 16 resource selection chains and the 8 task orders we have defined, we have 128 different rVNS methods.

Our first hyperheuristic, *HyperRandom*, selects at random a Low Level Heuristic (i.e. a (task order, resource selector) pair) to use at each iteration and applies it if its application will result in a positive improvement. This continues until no improvement has been found for a certain number of iterations. The second, *HyperGreedy*, evaluates all the Low Level Heuristics at each iteration and applies the best if it makes an improvement. This continues until no improvement is found. The low level heuristics are the combination of a task selector and a resource selector.

The genetic algorithm we will use is that of [3]. The chromosome represents an order of tasks to be scheduled by a serial scheduler. The initial population is generated randomly and the task order is evolved. The way in which the tasks are inserted into the schedule is a fast naïve approach as schedule must be generated many times per generation. The serial scheduler takes the next task from the chromosome and allocates resources to it greedily skill by skill. A resource is selected by finding the resource which has the greatest amount of available time in common with the task's time windows and any other resources already selected. After each skill has been allocated a resource, it is inserted into the schedule as early as possible. We use a population size of 50, mutation rate of 1%, and a crossover rate of 25% using Uniform Crossover based on out previous experience [3]. The GA is run for 100 generations (or for a maximum of 2.5 hours) and the result is the fittest individual in the final population.

4   Computational Experiments

To compare the methods for solving the problem, we use each method (one Genetic Algorithm, 128 rVNS and two hyperheuristics) on five different problem instances. The five problem instances require the scheduling of 400 tasks using 100 resources over one day using five different skills. Tasks require between one and three skills and resources possess between one and five skills. The problems are made to reflect realistic problems @Road Ltd. have identified and are generated using the problem generator used in [3].

Each method is used for five runs of the five instances and an average taken of the 25 results. To ensure fairness, each method is also run for a 2.5 hour "long-run" where the 25 results are repeatedly generated and the best average over all there repeated runs is reported. As these experiments require nearly a CPU year to complete (five runs of five instances using 131 different methods lasting 2.5 hours each = 8187.5 CPU hours) they were run in parallel on 60 identical 3.0 GHz Pentium 4 machines. Implementation was in C# .NET under Windows XP.

Figure 4 shows the results of the 2.5 hour "long run" for each rVNS approach. Results for a single run of each approach are similar but 1-4% worse on average. The intuitively "bad" task orders, *PriorityAsc* and *Random* are clearly shown to be worse than the intuitively reasonable orders such as *PriorityDesc*. Measure based on

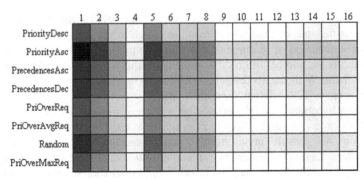

Fig. 4. Heat graph of the performance of rVNS methods for 2.5 hour "long run". Black = 4472, White =26525.

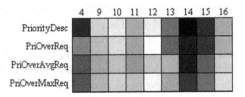

Fig. 4a. Heat graph of the performance of selected rVNS methods for 2.5 hour "long run". Black = 25398, White =26525.

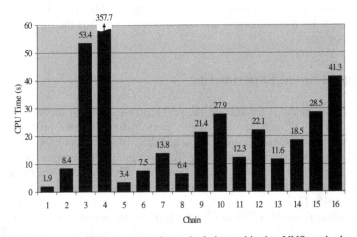

Fig. 5. Average CPU time taken by each chain used in the rVNS methods

decreasing priority or priority per hour (*PriorityDesc, PriOverReq, PriOverAvgReq, PriOverMaxReq*) are superior to other measures. Figure 4a compares the best approaches in detail. Chain 12 produces the best results for all task orders. It is clear to see the correlation between results with common chains or task orders. Chain 4 demonstrates that trying to estimate priority per hour is superior to *PriorityDesc*. This is probably because with a limited amount of free time in the schedule, using tasks that have lower priority but can be completed in a shorter time is more beneficial.

Figure 5 compares the CPU time for a single run of rVNS using each chain. It is clear that the approach would scale to very large problems using small resource selection sets such as for chains 1, 5, 6, 8, 11 and 13. Moreover, it appears from figure 4 that little solution quality is lost when covering the resources with small subsets rather than larger ones as in chain 4, but the CPU times are significantly reduced. Chain 12 yields the best results of the chains which take reasonable amounts of CPU time, and clearly outperform chain 2 and chain 7 which do not consider the whole set of resources. It seems that resources of poor competence must be considered to get the best possible results.

Table 2 shows the best result from the rVNS (Chain 12, Task Order *PriOverReq*) compared with GA and the hyperheuristic methods. They quite clearly show that *HyperGreedy* provided the fittest results on average while using more CPU time. The GA provided the worst result and in the slowest time. This may result from its insertion heuristic, however implementing a better one would make it even slower. The rVNS is the fastest method we have tested and provides results nearly 20% better than the GA in less than 1/350 of the CPU time required. Exactly solving small sub problems appears very effective in this case.

Table 2. GA, rVNS and Hyper-Heuristic Results for one run and long run

Method	Fitness (single run average)	CPU Time (s)	Fitness (after 2.5 hours)
GA	21401.3	9000.0	21401.3
rVNS (Best)	25662.5	25.1	26215.1
HyperRandom	24525.4	78.3	25645.4
HyperGreedy	26523.6	419.2	27103.1

HyperRandom performs poorly compared to the best rVNS method. rVNS task selectors and resource selectors are sensible guesses which significantly improve on the random approach of *HyperRandom*. The resource selectors of the rVNS tend to select resources which are of similar competence, so that a high competence resource is not combined with a low-competence resource (which might tie up the time of a high-competence resource).

The *HyperRandom*, and the *HyperGreedy* heuristics try significant numbers of bad low level heuristics which make local improvements which in the long run are far from optimal. In the case of the *HyperGreedy* method, the bad low level heuristics are evaluated every iteration which wastes CPU time. Analysis of the low level heuristics used in the *HyperGreedy* method was performed and show that 19 (26.4%) of the low level heuristics were never used and 56 (77.7%) of the low level heuristics were used less than one percent of the time. Figure 6 analyses the low level heuristics (LLHs) used. It shows the top 20 LLHs used together with when they are used in schedule generation. First third, middle third and last third show the usage at different stages in the scheduling process – from when the schedule is empty and unconstrained to when the schedule is almost full and inserting a task is more difficult. From these results it is clear that different LLHs contribute at different stages of the solution process, and that many different LLHs provide a contribution. For example, LLH 32 is more

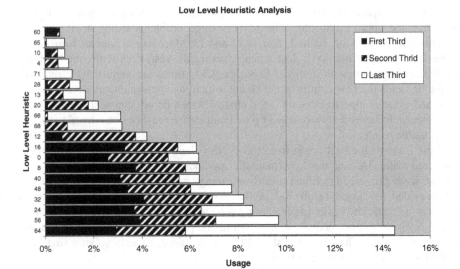

Fig. 6. Usage of low level heuristics throughout the HyperGreedy search

effective at the start of scheduling, LLH 12 is more effective in the middle and LLH 64 is more effective at the end. Without access to a large number of LLHs it seems that solution quality would be much reduced.

5   Conclusions

In this paper we have compared a large number (128) of reduced Variable Neighbourhood Search (rVNS) approaches to hyperheuristics and Genetic Algorithm approaches for workforce scheduling problem. We have demonstrated the effectiveness of heuristic/exact hybrids which find optimal subproblem solutions using an enumerative approach. Our rVNS method can produce good results to large problems in low CPU time. Our hyperheuristics produce even better results using more CPU time and we showed that the hyperheuristic uses a range of low level heuristics throughout the search process.

The hyperheuristics we used are simple and learning could potentially decrease CPU time and increase fitness. In future work we intend to implement learning mechanisms. We have seen from the analysis that many low level heuristics were never used and some used mainly at the beginning, middle or end. Learning the low level heuristics behaviour could potentially lead to better solutions in less time.

References

1. Hartmann, S.: Project Scheduling under Limited Resources: Model, methods and applications. Springer-Verlag, Berlin Heidelberg, New York (1999)
2. Pinedo, M. and Chao, X.: Operations scheduling with applications in manufacturing and services. McGraw-Hill, New York (1999)

3. Cowling, P., Colledge, N., Dahal, K. and Remde, S.: The Trade Off between Diversity and Quality for Multi-objective Workforce Scheduling. Evolutionary Computation in Combinatorial Optimization, Proc. Lecture Notes in Comp. Science 3906: 13-24 (2006)
4. Kolisch, R.: Serial and parallel resource-constrained project scheduling methods revisited: Theory and computation. European Journal of Oper. Res. 90 (2): 320-333 Apr 19 (1996)
5. Alcraz, J., Marotom R., and Ruiz, R.: Solving the Multi-mode Resource-Constrained Project Scheduling Problems with genetic algorithms. Journal of Operational Research Society (2004) 54, 614-626
6. Kolisch, R. and Hartmann, S.: Experimental Investigations of Heuristics for RCPSP: An Update. European Journal of Oper. Res. 174 (1): 23-37 (2006)
7. Bremermann, H.: The evolution of Intelligence. The Nervous System as a Model of it's environment. Technical Report No 1, contract No 477(17) Dept. of Math., Univ. of Washington, Seattle. (1958)
8. Holland, J. H.: Adaptation in Natural and Artificial Systems, Ann Abor, MI: University of Michigan Press, Michigan. (1975)
9. Whitley, D., Starkweather, T. and Shaner, D.: The travelling salesman and sequence scheduling: Quality solutions using genetic edge recombination. In Handbook of Genetic Algorithms, New York: Van Nostrand Reinhold (1991)
10. Falkenauer, E.: A Hybrid Grouping Genetic Algorithm for Bin Packing. Journal of Heuristics, vol 2, No. 1, 5-30 (1996)
11. Ross, P., Hart E. and Corne, D.: Some observations about GA-based exam timetabling. Lecture Notes in Computer Science 1408: 115-129 (1998)
12. Mladenovic, N. and Hansen, P.: Variable neighborhood search. Computers & Operational Research 24 (11): 1097-1100 Nov (1997)
13. Hansen, P. and Mladenovic, N.: Variable neighborhood search: Principles and applications. European Journal of Oper. Res. 130 (3): 449-467 May 1 (2001)
14. Fleszar, K. and Hindi, K.S.: Solving the resource-constrained project problem by a variable neighbourhood scheduling search. European Journal of Oper. Res. 155 (2): 402-413 Jun 1 (2004)
15. Garcia, C.G., Perez-Brito, D., Campos, V. and Marti, R.: Variable neighborhood search for the linear ordering problem. Comp. & Oper. Research 33 (12): 3549-3565 Dec (2006)
16. Sevkli, M., Aydin, M.E.: A variable neighbourhood search algorithm for job shop scheduling problems. Lecture Notes in Computer Science 3906: 261-271 (2006)
17. Cowling, P., Kendall, G. and Soubeiga, E.: A hyperheuristic approach to scheduling a sales summit. PATAT III Springer LNCS 2079: 176-190 (2001)
18. Fang, H., Ross, P. and Corne, D.: A Promising Hybrid GA/Heuristic Approach for Open-Shop Scheduling Problems. 11th European Conference on Artificial Intelligence, (1994)
19. Burke, E. K., Kendall, G., Soubeiga, E.: A tabu-search hyperheuristic for timetabling and rostering. Journal of Heuristics 9 (6): 451-470 Dec (2003)
20. Bai, R. and Kendall, G.: An Investigation of Automated Planograms Using a Simulated Annealing Based Hyper-heuristics. In proc. of The Fifth Metaheuristics Int. Conf. (2003)
21. Kendal, G., Han, L. and Cowling, P.: An Investigation of a Hyperheuristic Genetic Algorithm Applied to a Trainer Scheduling Problem. CEC, IEEE Press (2002) 1185-1190

An Analysis of Problem Difficulty for a Class of Optimisation Heuristics

Enda Ridge and Daniel Kudenko

Department of Computer Science, The University of York, York YO10 5DD, England
{ERidge,Kudenko}@cs.york.ac.uk

Abstract. This paper investigates the effect of the cost matrix standard deviation of Travelling Salesman Problem (TSP) instances on the performance of a class of combinatorial optimisation heuristics. Ant Colony Optimisation (ACO) is the class of heuristic investigated. Results demonstrate that for a given instance size, an increase in the standard deviation of the cost matrix of instances results in an increase in the difficulty of the instances. This implies that for ACO, it is insufficient to report results on problems classified only by problem size, as has been commonly done in most ACO research to date. Some description of the cost matrix distribution is also required when attempting to explain and predict the performance of these algorithms on the TSP.

1 Introduction and Motivation

Ant colony optimisation (ACO) algorithms [4] are a relatively new class of stochastic metaheuristic for typical Operations Research (OR) problems of combinatorial optimisation. To date, research has yielded important insights into ACO behaviour and its relation to other heuristics. However, there has been no rigorous study of the relationship between ACO algorithms and the difficulty of problem instances. Specifically, in ACO research to date, it has been mostly assumed that problem instance size is the main indicator of difficulty. Cheeseman *et al* [2] have shown that there is a relationship between the standard deviation of the cost matrix of a Travelling Salesperson Problem (TSP) instance and the difficulty of the problem for an *exact* algorithm.

In this paper we show that varying the standard deviation of a TSP instance cost matrix has a major impact on the difficulty of the problem instance for ACO algorithms. Consequently, we strongly recommend to take this cost matrix measure into account when evaluating and comparing ACO algorithms.

Our study focuses on Ant Colony System (ACS) [3] and Max-Min Ant System (MMAS) [9] since the field frequently cites these as its best performing algorithms. The study uses the TSP for the usual reasons of it being a popular abstraction of discrete combinatorial optimisation problems. Furthermore, the TSP has been instrumental in the development of ACO algorithms.

In our investigation we use established Design of Experiment (DOE) [8] techniques and statistical tools to explore data and test hypotheses. The designs and analyses from this paper could be applied to other stochastic heuristics for the

C. Cotta and J. van Hemert (Eds.): EvoCOP 2007, LNCS 4446, pp. 198–209, 2007.
© Springer-Verlag Berlin Heidelberg 2007

TSP. Problems are generated with a customised version of the freely available problem generator used in the DIMACS TSP Challenge[1]. The algorithms investigated are a Java port of the original C code that accompanies the book by Stützle and Dorigo [4][2]. This Java port has been informally verified to produce the same behaviour as the original.

The next Section gives a brief background on the ACO algorithms, ACS and MMAS. Section 3 describes the research methodology. Section 4 describes the results from the experiments. Related work is covered in Section 5. The paper ends with its conclusions and directions for future work.

2   Background

Ant Colony Optimisation algorithms are discrete combinatorial optimisation heuristics inspired by the foraging activities of natural ants. When applied to the TSP, the problem is represented by a graph of nodes and edges (representing the costs of visiting nodes). The objective is to minimise the cost of visiting all nodes in the graph once and only once. This abstraction has application in problems of traffic routing and manufacture among others.

Broadly, the ACO algorithms work by placing a set of artificial ants on the TSP nodes. The ants build TSP solutions by moving between nodes along the graph edges. These movements are probabilistic and are influenced both by a heuristic function and the levels of a real-valued marker called a *pheromone*. Their movement decisions also favour nodes that are part of a candidate list, a list of the least costly cities from a given node. The iterated activities of artificial ants lead to some combinations of edges becoming more reinforced with pheromone than others. Eventually the ants converge on a solution.

It is common practice to hybridise ACO algorithms with local search procedures. This study focuses on ACS and MMAS as constructive heuristics and so omits any such procedure. Adding one of a potentially infinite number of local search variants confounds any effects on the ant algorithm with effects on the local search component. We see this confounding in much of the ACO literature where effects that may be due to an interaction with the local search procedure are attributed to an interaction with the ACO algorithm. The interested reader is referred to the most recent review text [4] for further information on ACO algorithms.

In common with many heuristics, ACO algorithms have a large number of tuning parameters that can have a dramatic effect on algorithm behaviour. The parameters used in this paper are listed in Table 1.

We chose these values because they are commonly listed in the field's main book [4] and literature. We stress that we by no means support such a 'folk' approach to parameter selection in general. Our ultimate goal is to select parameter values methodically using RSM techniques. This study is a prerequisite step towards that goal. In this paper's context, selecting parameter values as we

---

[1] http://www.research.att.com/~dsj/chtsp/

[2] http://iridia.ulb.ac.be/~mdorigo/ACO/aco-code/public-software.html

Table 1. Parameter settings for the ACS and MMAS algorithms. Value are taken from the original publications [4,9]. Please refer to these for a detailed explanation of the parameters.

Parameter	Symbol	ACS	MMAS
Ants	m	10	25
Pheromone emphasis	α	1	1
Heuristic emphasis	β	2	2
Candidate List length		15	20
Exploration threshold	q_0	0.9	N/A
Pheromone decay	ρ_{global}	0.1	0.8
Pheromone decay	ρ_{local}	0.1	N/A
Ant activity		Sequential	Sequential

have done shows that we did not contrive a result by searching for a unique set of values that would demonstrate our desired effect. Furthermore, it makes our conclusions with freely available code applicable to all research that has used these parameter values without justification. For the purposes of this paper, demonstrating an effect of cost matrix standard deviation on performance with even one parameter set is sufficient to merit this factor's consideration in all related studies.

3　Method

This section describes the general experiment design issues relevant to this paper. Others have covered these in detail [7] for heuristics in general. Further information on Design of Experiments is available in the literature [8].

3.1　Stopping Criterion

The choice of stopping criterion for an experiment run is difficult when algorithms can continue to run and improve indefinitely. CPU time is certainly not a scientifically reproducible metric and some independent metric such as a combinatorial count of an algorithm operation is often used.

A problem with this approach is that our choice of combinatorial count can bias our results. Should we stop after 1000 iterations or 1001? We mitigate this concern by taking 10 evenly spaced measurements over 5000 iterations of the algorithms and separately analysing the data at all 10 measurements. Note that a more formal detection of possible differences introduced by different stopping criteria would have required a different analysis.

3.2   Response

We measure the percentage relative error from the known optimum solution. Other solution quality measures have been proposed, notably the *adjusted differential approximation* [11]. We did not investigate such measures here. Concorde [1] was used to calculate the optima of the generated instances. To make the data amenable to statistical analysis, a transformation was required. The response was transformed using either a $\log_{10}$ or inverse square root transformation as recommended by the Box-Cox plot technique.

3.3   Outliers

An outlier is a data value that is either unusually large or unusually small relative to the rest of the data. Outliers are important because their presence can distort data and render statistical analyses inaccurate. There are several approaches to dealing with outliers. This research used the approach of deleting outliers from an analysis until the analysis passed the usual diagnostics mentioned in Section 4.1.

3.4   Experiment Design

This study uses a *two-stage nested* (or *hierarchical*) design. Consider this analogy.

A company receives stock from several suppliers. They test the quality of this stock by taking 10 samples from each supplier's batch. They wish to determine whether there is a significant overall difference between supplier quality and whether there is a significant quality difference in samples within a supplier's batch. A full factorial design of the supplier and sample factors is inappropriate because samples are unique to their supplier.

A similar situation arises in this research. An algorithm encounters TSP instances with different levels of standard deviation of cost matrix. We want to determine whether there is a significant overall difference in algorithm solution quality for different levels of standard deviation. We also want to determine whether there is a significant difference in algorithm quality between instances that have the same standard deviation. Figure 1 illustrates the two-stage nested design schematically.

The standard deviation of the generated instance is the parent factor and the individual instance number is the nested factor. Therefore, an individual treatment consists of running the algorithm on a particular instance generated with a particular standard deviation. This design applies to an instance of a given size and therefore cannot capture possible interactions between instance size and instance standard deviation. Capturing such interactions would require a more complicated *crossed nested design*. This research uses the simpler design

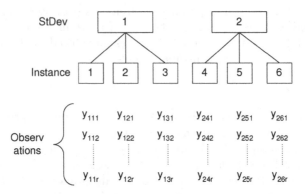

Fig. 1. Schematic for the Two-Stage Nested Design with r replicates. (adapted from [8]). Note the instance numbering to emphasise the uniqueness of instances within a given level of standard deviaiton.

to demonstrate that standard deviation is important for a given size. Interactions are captured in the designs mentioned in Section 7.

Because of the stochastic nature of ACS and MMAS, we replicated each treatment run 10 times with a different random seed. Available computational resources necessitated running experiments across a variety of similar machines. Runs were executed in a randomised order across these machines to counteract any uncontrollable nuisance factors.

3.5 Instances

Problem instances were created using a modification of the portmgen generator from the DIMACS TSP challenge. The original portmgen created a cost matrix by choosing costs uniformly randomly within a certain range. We first ported this generator into Java and verified that its behaviour was unchanged. We then adjusted the generator so that edge costs could be drawn from any distribution. In particular, we followed Cheeseman *et al*'s [2] approach and drew edge costs from a Log-Normal distribution. Although Cheeseman *et al* did not state their motivation for using such a distribution, a plot of the relative frequencies of the normalised edge costs of instances from a popular online benchmark library, TSPLIB, shows that the majority have a Log-Normal shape (Figure 2).

An appropriate choice of inputs results in a Log-Normal distribution with a desired mean and standard deviation. We created instances of a given size with a mean fixed at 100. Standard deviation was varied across 5 levels: 10, 30, 50, 70 and 100. Figure 3 shows relative frequencies of the normalised cost matrices of several generated instances.

Three problem sizes; 300, 500 and 700 were tested. The same instances were used for all algorithms.

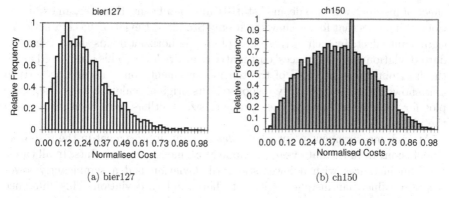

(a) bier127   (b) ch150

Fig. 2. Cost matrix distribution of two TSPLIB instances

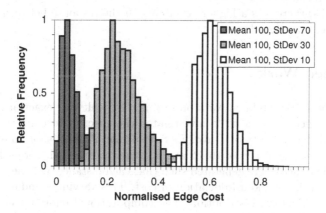

Fig. 3. Relative frequencies of normalised edge costs for several instances of the same size and same mean cost. Instances are distinguished by their standard deviation.

4   Results and Analysis

4.1   Analysis Procedure

The two-stage nested designs were analysed with the General Linear Model. Standard deviation was treated as a fixed factor since we explicitly chose its levels and instance was treated as a random factor. The usual diagnostic tools were used to verify that the model was correct and that its assumptions had not been violated—model fit, normality, constant variance, time-dependent effects, and leverage. Further details on these analyses and diagnostics are available in many textbooks [8].

4.2   Results

Figure 4 illustrates the results. In all cases, the effect of Standard Deviation on solution quality was deemed statistically significant at the $p < 0.01$ level. The effect of instance was also deemed statistically significant. However, an examination of the plots that follow shows that only Standard Deviation has a practically significant effect. In each of these box-plots, the horizontal axis shows the standard deviation of the instances' cost matrices at five levels. This is repeated along the horizontal axis at three of the ten measurement points used. The vertical axis shows the solution quality response in its original scale. There is a separate plot for each algorithm and each problem size. Outliers have been included in the plots.

At each measurement point, there was a slight improvement in the response. Conclusions from the data were the same at all measurement points. In all cases, problem instances with a lower standard deviation had a significantly lower response value than instances with a higher standard deviation. This difference was greatest between instances with a standard deviation of 10 and those with a standard deviation of 30.

In all cases, there was a higher variability in the response between instances with a higher standard deviation.

5   Related Work

There has been some related work on problem difficulty for exact and heuristic algorithms. Cheeseman *et al* [2] investigated the effect of cost matrix standard deviation on the difficulty of Travelling Salesperson Problems for an exact algorithm. Three problem sizes of 16, 32 and 48 were investigated. For each problem size, many instances were generated such that each instance had the same mean cost but a varying standard deviation of cost. This varying standard deviation followed a Log-Normal distribution. The computational effort for an exact algorithm to solve each of these instances was measured and plotted against the standard deviation of cost matrix. This paper differs from Cheeseman *et al* in that it uses larger problem sizes and a heuristic algorithm rather than exact algorithm. Furthermore, its conclusions are reinforced with a DOE approach and statistical analyses.

Fischer *et al* [5] investigated the influence of Euclidean TSP structure on the performance of two algorithms, one exact and one approximate. The former was branch-and-cut [1] and the latter was the iterated Lin-Kernighan algorithm [6]. In particular, the TSP structural characteristic investigated was the distribution of cities in Euclidean space. The authors varied this distribution by taking a structured problem instance and applying an increasing perturbation to the city distribution until the instance resembled a randomly distributed problem. There were two perturbation operators. A reduction operator removed between 1% to 75% of the cities in the original instance. A shake operator offset cities from their original location. Using 16 original instances, 100 perturbed instances were created for each of 8 levels of the perturbation factor. Performance on perturbed instances was compared to 100 instances created by uniformly randomly distributing cities in a square. Predictably, increased perturbation lead to increased solution times that were closer to the times for a completely random

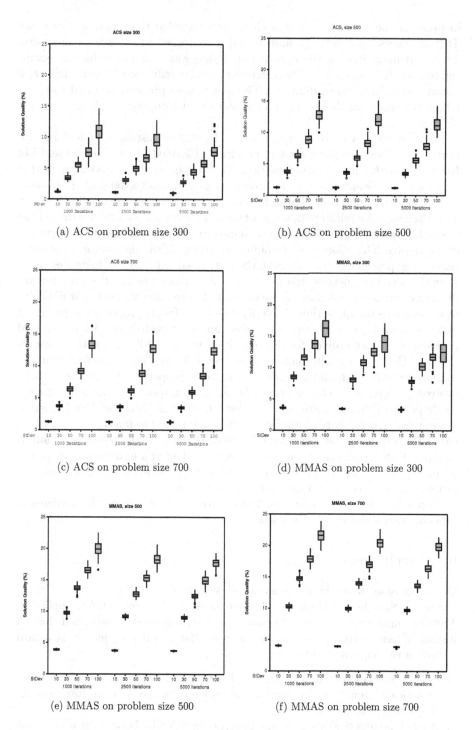

(a) ACS on problem size 300

(b) ACS on problem size 500

(c) ACS on problem size 700

(d) MMAS on problem size 300

(e) MMAS on problem size 500

(f) MMAS on problem size 700

Fig. 4. Boxplots of results for both algorithms and all problem sizes

instance of the same size. It was therefore concluded that random Euclidean TSP instances are relatively hard to solve compared to structured instances. An unfortunate flaw in the experiment design was that the reduction operator confounded changed problem structure with a reduction in problem size, a known factor in problem difficulty. This paper fixes problem size and cost matrix mean and controls cost matrix standard deviation, thus avoiding any such confounding.

Most recently, Van Hemert [10] evolved problem instances of a fixed size that were difficult to solve for two heuristics: Chained Lin-Kernighan and Lin Kernighan with Cluster Compensation. TSP instances of size 100 were created by uniform randomly selecting 100 coordinates from a 400x400 grid. This seems to be a similar generation approach to the **portgen** generator from the DIMACS TSP challenge. An initial population of such instances was evolved for each of the algorithms where higher fitness was assigned to instances that required a greater effort to solve. This effort was a combinatorial count of the algorithms' most time-consuming procedure. Van Hemert then analysed the evolved instances using several interesting metrics. His aim was to determine whether the evolutionary procedure made the instances more difficult to solve and whether that difficulty was specific to the algorithm. To verify whether difficult properties were shared between algorithms, each algorithm was run on the other algorithm's evolved problem set. A set evolved for one algorithm was less difficult for the other algorithm. However, the alternative evolved set still required more effort than the random set indicating that some difficult instance properties were shared by both evolved problem sets. Our approach began with a specific hypothesis about a single problem characteristic and its effect on problem hardness. Van Hemert's, by contrast, evolved hard instances and then attempted to infer, *post-hoc*, which characteristics might be responsible for that hardness. If the researcher's aim is to stress test a heuristic, then we believe Van Hemert's approach is more appropriate. The approach presented here is appropriate when isolating a specific problem characteristic that may affect problem hardness.

To our knowledge, these are the first rigorous experiments on the hardness of problem instances for ACO heuristics.

6   Conclusions

Our conclusions from the aforementioned results are as follows.

For the Stützle and Dorigo implementations of ACS and MMAS, applied to TSP instances generated with log-normally distributed edge costs such that all instances have a fixed cost matrix mean of 100 and a cost matrix standard deviation varying from 10 to 100:

1. an increase in cost matrix standard deviation leads to a statistically and practically significant increase in the difficulty of the problem instances for these algorithms.
2. there is no practically significant difference in difficulty between instances that have the same size, cost matrix mean and cost matrix standard deviation.

3. there is no practically significant difference between the difficulty measured after 1000 algorithm iterations and 5000 algorithm iterations.

These results are important for the ACO community for the following reasons:

- They demonstrate in a rigorous, designed experiment fashion, that quality of solution of an ACO TSP algorithm is affected by the standard deviation of the cost matrix.
- They demonstrate that cost matrix standard deviation must be considered as a factor when building predictive models of ACO TSP algorithm performance.
- They clearly show that performance analysis papers using ACO TSP algorithms must report instance cost matrix standard deviation as well as instance size since two instances with the same size can differ significantly in difficulty.
- They motivate an improvement in benchmarks libraries so that they provide a wider crossing of both instance size and instance cost matrix standard deviation. Plots of instances in the TSPLIB show that *generated* instances generally have the same shaped distribution of edge costs. However, *real* instances often have different shaped distributions

For completeness and for clarity, we state that this research does *not* examine the following issues.

- We are not examining clustered problem instances or grid problem instances. These are other common forms of TSP in which nodes appear in clusters and in a very structured grid pattern respectively.
- Algorithm performance is not being examined since no claim was made about the suitability of the parameter values used. Rather, we aim to demonstrate an effect for standard deviation and so argue that its should be included as a factor in experiments that do examine algorithm performance.
- We make no direct comparison between algorithms since algorithms were not tuned methodically. That is, we are not entitled to say that ACS did better than MMAS on, say, instance X with a standard deviation of Y.
- We make no direct comparison of the response values for different sized instances. Clearly, 3000 iterations explores a bigger fraction of the search space for 300-city problems than for 500 city problems. Such a comparison *could* be made if it was clear how to scale iterations with problem size. Such scaling is an open question.

7   Future Work

There are several avenues of future work leading from this paper.

The same analysis is worthwhile for other popular ACO algorithms. The code provided by Stützle and Dorigo also has implementations of Best-Worst Ant System, Ant System, Rank-Based Ant System and Elitist Ant System. We are conducting these analyses presently.

One of the main motivations of this paper's research was to investigate whether inclusion of cost matrix standard deviation as a factor in experiment designs improves the predictive capabilities of Response Surface Models. Recall that we have used fixed algorithm parameter settings from the literature. Screening experiments prior to Response Surface design would establish which if any of these parameters interacts with cost matrix standard deviation to affect performance. Conceivably, for example, the number of ants used by the algorithm might mitigate the effects of the cost matrix standard deviation. Until this is established in a methodical and reproducible fashion, we cannot ignore the possible influence of this factor in our experiments with ACO algorithms.

The use of a well-established Design of Experiments approach with two-stage nested designs and analysis with the General Linear Model could also be applied to other heuristics for the TSP. It is important that we introduce such rigour into the field so that we can move from the competitive testing of highly engineered designs to the scientific evaluation of hypotheses about algorithms.

Acknowledgements

The authors are grateful to the reviewers for the helpful and encouraging comments and criticisms.

References

1. D. Applegate, R. Bixby, V. Chvatal, and W. Cook. Implementing the Dantzig-Fulkerson-Johnson algorithm for large traveling salesman problems. *Mathematical Programming Series B*, 97(1-2):91–153, 2003.
2. Peter Cheeseman, Bob Kanefsky, and William M. Taylor. Where the Really Hard Problems Are. In *Proceedings of the Twelfth International Conference on Artificial Intelligence*, volume 1, pages 331–337. Morgan Kaufmann Publishers, Inc., USA, 1991.
3. Marco Dorigo and Luca Maria Gambardella. Ant Colony System: A Cooperative Learning Approach to the Traveling Salesman Problem. *IEEE Transactions on Evolutionary Computation*, 1(1):53–66, 1997.
4. Marco Dorigo and Thomas Stützle. *Ant Colony Optimization*. The MIT Press, Massachusetts, USA, 2004.
5. Thomas Fischer, Thomas Stützle, Holger Hoos, and Peter Merz. An Analysis Of The Hardness Of TSP Instances For Two High Performance Algorithms. In *Proceedings of the Sixth Metaheuristics International Conference*, pages 361–367. 2005.
6. K. Helsgaun. An effective implementation of the Lin-Kernighan traveling salesman heuristic. *European Journal of Operational Research*, 126(1):106–130, 2000.
7. David S. Johnson. A Theoretician's Guide to the Experimental Analysis of Algorithms. In Goldwasser, Johnson, and McGeoch, editors, *Proceedings of the Fifth and Sixth DIMACS Implementation Challenges*, pages 215–250. American Mathematical Society, 2002.
8. Douglas C. Montgomery. *Design and Analysis of Experiments*. John Wiley and Sons Inc, 6 edition, 2005.

9. Thomas Stützle and Holger H. Hoos. Max-Min Ant System. *Future Generation Computer Systems*, 16(8):889–914, 2000.
10. Jano I. van Hemert. Property Analysis of Symmetric Travelling Salesman Problem Instances Acquired Through Evolution. In Günther R. Raidl and Jens Gottlieb, editors, *Proceedings of the Fifth Conference on Evolutionary Computation in Combinatorial Optimization*, volume 3448 of *LNCS*, pages 122–131. Springer-Verlag, Berlin, 2005.
11. M. Zlochin and Marco Dorigo. Model based search for combinatorial optimization: a comparative study. In J. J. M. Guervs, P. Adamidis, and H.-G. Beyer, editors, *Proceedings of the Seventh International Conference on Parallel Problem Solving from Nature*, volume 2439, pages 651–661. Springer-Verlag, Berlin, Germany, 2002.

A New Grouping Genetic Algorithm for the Quadratic Multiple Knapsack Problem

Alok Singh[1] and Anurag Singh Baghel[2]

[1] J. K. Institute of Applied Physics and Technology, Faculty of Science,
University of Allahabad, Allahabad – 211002, UP, India
alok@jkinstitute.org
[2] Department of Electronics and Communication, Banasthali Vidyapith Jaipur Campus,
Sarojini Marg, Jaipur – 302001, Rajasthan, India
anuragsbaghel@yahoo.com

Abstract. The quadratic multiple knapsack problem is an extension of the well known 0/1 multiple knapsack problem. In the quadratic multiple knapsack problem, profit values are associated not only with individual objects but also with pairs of objects. Profit value associated with a pair of objects is added to the overall profit if both objects of the pair belong to the same knapsack. Being an extension of the 0/1 multiple knapsack problem, this problem is also NP-Hard. In this paper, we have proposed a new steady-state grouping genetic algorithm for the quadratic multiple knapsack problem and compared our results with two recently proposed methods – a genetic algorithm and a stochastic hill climber. The results show the effectiveness of our approach.

Keywords: Combinatorial optimization, grouping genetic algorithm, knapsack problem, quadratic multiple knapsack problem.

1 Introduction

The quadratic multiple knapsack problem (QMKP) is an extension of the well known 0/1 multiple knapsack problem (MKP). In MKP we are given a set of n objects, and K knapsacks. Each object i, $i \in \{1, 2, ..., n\}$ has profit p_i and weight w_i. Each knapsack j, $j \in \{1, 2, ..., K\}$ has a capacity C_j. The multiple knapsack problem consists in selecting the K disjoint subsets of objects to be put into the K knapsacks such that the total weight of objects in subset j should not exceed C_j and the overall profit of all the selected objects is as large as possible. By introducing binary variables x_{ij} to indicate whether object i is included in knapsack j ($x_{ij} = 1$) or not ($x_{ij} = 0$), the MKP can be formulated as:

Maximize
$$P = \sum_{i=1}^{n} \sum_{j=1}^{K} x_{ij} p_i$$

Subject to

$$\sum_{i=1}^{n} x_{ij} w_i \le C_j, \quad j = 1, 2, ... K$$

C. Cotta and J. van Hemert (Eds.): EvoCOP 2007, LNCS 4446, pp. 210–218, 2007.
© Springer-Verlag Berlin Heidelberg 2007

$$\sum_{j=1}^{K} x_{ij} \leq 1, \quad i = 1, 2, ..., n$$

$$x_{ij} \in \{0,1\} \quad i = 1, 2, ..., n \quad j = 1, 2, ..., K$$

MKP is NP-Hard as for $K=1$ it reduces to 0/1 Knapsack Problem [1]. QMKP is an extension of MKP. The only difference between QMKP and MKP is that in QMKP profit values are associated not only with individual objects but also with pairs of objects. The profit value p_{ij} associated with a pair of objects i and j is added to the over all profit, if both object i and j belong to the same knapsack. Therefore QMKP seeks to maximize

$$P = \sum_{i=1}^{n} \sum_{j=1}^{K} x_{ij} p_i + \sum_{i=1}^{n-1} \sum_{j=i+1}^{n} \sum_{k=1}^{K} x_{ik} x_{jk} p_{ij}$$

QMKP is NP-Hard as it reduces to MKP when all p_{ij} are zero.

Though MKP is widely studied and a number of evolutionary algorithms have been proposed for it [2, 3, 4], QMKP is only recently defined and studied by Hiley and Julstrom [5]. They considered a restricted version of QMKP in which all capacities C_j are same. They presented three methods – a greedy heuristic, a generational genetic algorithm and a stochastic hill climber for their version of QMKP. On the test instances considered, stochastic hill climber obtained better solution on average followed by genetic algorithm and greedy heuristic. Genetic algorithm performs better on instances with small K, but its performance decline sharply as K grows. Hereafter this genetic algorithm will be referred to as HJ-GA and stochastic hill-climber as HJ-SHC. Hiley and Julstrom [5] cited a practical application of QMKP in the situation where a manager has to select persons for multiple projects, each with its own budget, at the same time. The manager knows the salary of each person and, the productivity of each person, both individually and in pairs. Obviously, the manager will try to assign persons to projects in such a way that maximizes the overall productivity without exceeding the budget of any project.

Clearly QMKP is a grouping problem [6, 7], i.e. a problem that seeks an optimal assignment of objects according to a given cost function into different groups subject to some constraints. Therefore when designing a genetic algorithm for this problem, genetic operators should be designed in such a way that these operators should try to preserve grouping information as far as possible while generating new chromosomes [6, 7]. The genetic algorithm for the QMKP that this paper describes is designed with exactly the aforementioned idea. Falkenauer [6, 7] named such type of genetic algorithms as grouping genetic algorithms. Like Hiley and Julstrom [5], we also assume that all knapsack capacities C_j are same and are equal to C. We have compared our genetic algorithm with the genetic algorithm and the stochastic hill climber proposed by Hiley and Julstrom [5]. The results show the effectiveness of our approach.

This paper is organized as follows: Section 2 describes our grouping genetic algorithm. Computational results are presented in section 3, whereas section 4 outlines some conclusions.

2 The Grouping Genetic Algorithm

We have developed a steady-state grouping genetic algorithm (SSGGA) for the QMKP. Steady-state genetic algorithm uses steady-state population replacement method [8]. In this method genetic algorithm repeatedly selects two parents, performs crossover and mutation to produce a single child that replaces a less fit member of the population. This is different from generational replacement, where a new population of children is created and the whole parent population is replaced. The steady-state population replacement method has an advantage over generational method due to the fact that the best solutions are always kept in the population and the child is immediately available for selection and reproduction. Thus we can possibly find better solutions quicker. Moreover with steady-state population replacement method we can easily avoid the multiple copies of the same individual in the population. In the generational approach multiple copies of the same individual can exist in the population. Though these individuals are usually the best individuals, they can rapidly dominate the whole population. In this situation, no further improvement in solution quality is possible without mutation, and often, a much higher mutation rate is required to get further improvements. In the steady-state approach the child can be easily checked against the existing population members and if it is identical to any existing individual in the population then it is discarded. In this way the problem of premature convergence is deterred by disallowing the multiple copies of the same individual in the population. The main features of SSGGA are described below. However, before describing the main features of SSGGA, we need to define the concept of relative value density [5]. The relative value density of an object i with respect to a set S of objects is the sum of its profit value p_i and all profit values p_{ij} such that $j \in S$ divided by its weight.

2.1 Chromosome Representation

Chromosome is represented as set of knapsacks i.e. there is no ordering among the knapsacks. With such a representation there is no redundancy. Every solution is represented uniquely.

2.2 Fitness

Fitness of a chromosome is equal to the overall profit of the solution it represents.

2.3 Crossover

Our crossover operator is derived from the crossover operator proposed in [9] for the one-dimensional bin-packing problem. Our crossover operator consists of two phases. First phase iteratively builds the child chromosome. During each iteration, it selects one of the two parents uniformly at random and copies the knapsack with largest profit value from the selected parent to the child. Then it deletes all the objects belonging to this knapsack from both the parents and profit values of the knapsacks of the parents are updated accordingly. This process is repeated K times. The second phase iteratively tries to include as many unassigned objects as it can into the knapsacks without violating the capacity constraints. During each iteration it selects a knapsack at random and adds to it the unassigned object that fits and has highest relative value density with respect to the objects already in the selected knapsack. This process is repeated until it becomes impossible to add any more objects to any of the knapsacks.

We have used binary tournament selection to select the two parents for crossover, where more fit candidate is selected with probability p_{better}.

Similar to [5, 9] here also crossover and mutation is used in a mutually exclusive manner, i.e. each child is generated by either the crossover operator or the mutation operator but never by both. Crossover is applied with probability p_c, otherwise mutation is used.

2.4 Mutation

The mutation operator removes some of the objects from knapsacks. Then it proceeds similar to the second phase of crossover operator. We have used 3-ary tournament selection to select a chromosome for mutation, where the candidate with better fitness is selected with probability 1.0. 3-ary tournament selection is used because more fit chromosome has greater chance of generating a better chromosome after mutation.

2.5 Replacement Policy

The generated child is first tested for uniqueness against the existing population members. If it is unique then it always replaces the least fit member of the population, otherwise it is discarded.

2.6 Initial Population Generation

Each member of the initial population is generated using a procedure that is derived from greedy heuristic proposed in [5]. The procedure used here differs from the greedy heuristic in that it selects first object of each knapsack randomly from the list of unassigned objects rather than the unassigned object with highest relative density with respect to the list of unassigned objects. Initially all the knapsacks are empty. The procedure fills each knapsack one by one. The first object of the knapsack is selected randomly as already described. Then objects are added to the knapsack iteratively. During each iteration, an unassigned object that fits and has highest relative value density with respect to the objects already in the knapsack is added to the knapsack. This process is repeated until it becomes impossible to add any more objects to the knapsack. After this the next knapsack is filled in the same way. The whole process is repeated until all the knapsacks have been filled.

Each newly generated chromosome is checked for uniqueness against the population members generated so far and if it is unique it is included in the initial population otherwise it is discarded.

3 Computational Results

SSGGA has been coded in C and executed on a Pentium 4 system with 512 MB RAM, running at 2.4 GHz under Red-Hat Linux 9.0. In all our experiments with SSGGA we have used $p_c = 0.6$, $p_{better} = 0.8$. Mutation try to remove each object allocated to a knapsack with probability ($2 \times K/nobj$), where $nobj$ is the number of objects allocated to knapsacks. With this probability mutation will remove on an average two objects per knapsack. The mutation operator of HJ-GA [5] also removes two objects from each knapsack. The population size of SSGGA is equal to n, the number of objects in the test instance. All the parameter values were chosen after large number of trials. These parameter values provide good results, although these values are in no way optimal for all problem instances. We have tested SSGGA on the same 60 QMKP instances as used by Hiley and Julstrom [5]. These instances are characterized by three things – the density d (proportion of non-zero p_{ij}), number of objects n, and the number of knapsacks K. For every instance, the knapsack capacities are set to 80% of the sum of instance's object's weights divided by K. For these instances d is either 0.25 or 0.75, n is either 100 or 200 and K can take any value from {3, 5, 10}. There are 5 instances for a particular d, n and K, resulting in a total of 60 instances. Originally these instances are the instances of the quadratic knapsack problem and are available at *http://cermsem.univ-paris1.fr/soutif/QKP/QKP.html*. SSGGA was executed 40 times on each instance, each time with a different random seed. During each run both HJ-GA and HJ-SHC generate 20000 candidate solutions. Therefore to allow a fair comparison with HJ-GA and HJ-SHC, SSGGA also generates 20000 candidate solutions.

Tables 1 and 2 compare the performance of SSGA with HJ-GA and HJ-SHC. Table 1 reports the performance of three algorithms on instances with $d = 0.25$, whereas table 2 reports the same for $d = 0.75$. Data for HJ-GA and HJ-SHC are taken from [5]. For each instance, tables 1 and 2 report the number of objects n, the number of knapsacks K in it, its number and knapsacks capacity C. For each of the three approaches on each instance the tables report the best and average value of the solution, the standard deviation of solution values and average execution time in seconds.

Tables 1 and 2 clearly show the superiority of SSGGA over HJ-GA and HJ-SHC. Average solution values obtained by SSGGA are always better than those obtained with HJ-GA and HJ-SHC. In comparison to HJ-GA, the best solution of SSGGA is better on 56 instances and worse on 4 instances. The best solution of SSGGA is better than that of HJ-RHC on 57 instances and worse on 3 instances.

HJ-GA and HJ-SHC were executed on Pentium 4, 2.53 GHz system with 1 GB RAM, whereas SSGGA was executed on Pentium 4, 2.4 GHz system with 512 MB RAM. Therefore it is not possible to exactly compare the running times of SSGGA with those of HJ-GA and HJ-SHC. However, on instances with $K = 3$, SSGGA is clearly slower to HJ-GA and HJ-SHC, whereas on instances with $K = 10$, it is clearly faster. On instances with $K = 5$, running times of SSGGA are less in comparison to HJ-SHC and roughly the same in comparison to HJ-GA. The running time of HJ-GA and HJ-SHC increases with increase in K, whereas that of SSGGA decreases. The

Table 1. Performance of HJ-SHC, HJ-GA and SSGGA on instances with d = 0.25

Instance				HJ-SHC				HJ-GA				SSGGA			
n	K	Num	C	Best	Avg.	SD	Time[a]	Best	Avg.	SD	Time[a]	Best	Avg.	SD	Time
100	3	1	688	28144	27635	294	1.49	28665	27904	339	0.97	28798	28485	195	1.78
		2	738	29915	26222	377	1.61	28059	27044	421	1.00	28036	27507	247	1.75
		3	663	25945	25193	365	1.45	26780	25991	344	0.95	26936	26411	283	1.75
		4	804	27109	26127	431	1.67	28199	27265	497	1.02	28418	27473	483	1.77
		5	723	26288	25617	297	1.58	27550	26683	397	1.00	27617	26971	295	1.72
100	5	1	413	21584	20911	289	2.00	21914	21315	316	1.20	22038	21662	191	1.27
		2	442	20934	19768	322	2.11	21216	20472	326	1.25	21459	21046	199	1.30
		3	398	19454	18765	255	1.99	20243	19763	296	1.17	21012	20279	495	1.24
		4	482	20173	19730	279	2.30	21698	20923	291	1.31	21987	21344	409	1.26
		5	434	19932	18843	266	2.12	20808	20248	259	1.25	21057	20304	370	1.25
100	10	1	206	15232	14737	240	3.86	13521	12499	419	2.19	15663	15201	177	1.49
		2	221	14210	13684	243	4.11	12859	12019	299	2.31	15002	14654	207	1.44
		3	199	13334	12918	196	3.88	11790	11245	278	2.15	14231	13716	220	1.41
		4	241	14321	13867	225	4.36	13316	12593	333	2.44	15979	15310	351	1.43
		5	217	13405	12929	210	4.14	11909	11389	269	2.25	14510	14018	225	1.45
200	3	1	1381	99232	98169	548	4.54	97469	95497	991	3.24	99753	99286	289	5.29
		2	1246	106730	105857	469	4.38	106162	100521	3242	3.13	107475	107036	210	5.46
		3	1335	103529	102475	505	4.56	101291	97157	2099	3.21	103607	102952	254	5.36
		4	1413	97407	97067	831	4.55	95649	93968	812	3.29	98276	97092	477	5.14
		5	1358	100827	99762	628	4.46	99458	96077	1815	3.25	101463	100612	294	5.39
200	5	1	828	72277	70776	593	5.77	70731	68705	974	3.70	73040	72216	372	3.41
		2	747	77551	76643	502	5.38	76297	72924	1200	3.51	78428	77236	404	3.66
		3	801	75409	74497	594	5.60	74377	72924	2050	3.69	76321	75835	212	3.55
		4	848	71307	69417	612	5.87	70264	67416	1138	3.83	71964	70892	399	3.40
		5	815	74287	73229	465	5.79	72745	69978	1439	3.72	74936	74538	173	3.54
200	10	1	414	48006	46960	609	10.16	42016	39791	982	5.77	49212	48065	426	3.03
		2	373	51438	50622	404	9.09	45483	42739	1303	5.38	52153	51568	238	3.09
		3	400	50717	49688	480	9.89	45698	42475	1861	5.75	51205	50694	231	3.10
		4	424	47296	45751	645	10.15	41623	39446	963	5.94	47853	47001	323	3.06
		5	407	50402	49431	514	9.84	46811	42399	2023	5.83	51000	50267	293	3.04

a – Execution time on a Pentium 4, 2.53 GHz system with 1 GB RAM

Table 2. Performance of HJ-SHC, HJ-GA and SSGGA on instances with $d = 0.75$

n	K	Num	C	HJ-SHC Best	Avg.	SD	Time[a]	HJ-GA Best	Avg.	SD	Time[a]	SSGGA Best	Avg.	SD	Time
100	3	1	699	69786	69172	327	1.36	69769	68941	480	1.01	69935	69694	133	1.89
		2	714	69056	68508	313	1.41	69146	68639	340	1.03	69344	69203	71	1.75
		3	686	68547	67939	361	1.46	68763	67557	832	1.02	68776	68518	199	1.76
		4	666	69646	69003	437	1.46	69907	69101	861	1.04	69696	69677	40	2.05
		5	668	69480	68578	341	1.44	69410	68856	306	1.02	69520	69262	146	1.70
100	5	1	401	48888	48138	286	1.97	48663	47678	749	1.26	48675	48414	125	1.35
		2	428	48686	48028	317	1.95	48990	48175	398	1.32	48916	48376	249	1.27
		3	411	47396	46970	244	1.89	47512	46623	503	1.29	48126	47815	156	1.32
		4	400	49468	48864	254	1.94	49845	49194	295	1.30	49724	49297	150	1.34
		5	400	47982	47298	311	1.90	47925	47230	554	1.29	48746	48148	267	1.31
100	10	1	200	29136	28665	262	3.56	26603	25681	688	2.23	29179	28762	136	1.39
		2	214	30367	30031	187	3.86	28663	27815	391	2.43	30640	30357	120	1.42
		3	205	28838	28297	238	3.76	26176	25038	562	2.31	28857	28561	152	1.34
		4	200	30624	30346	199	3.89	29701	28592	333	2.40	31039	30581	238	1.40
		5	200	29375	28956	206	3.70	27130	25937	650	2.28	29641	29240	147	1.40
200	3	1	1311	269447	267765	809	4.33	268919	265523	1820	3.40	269351	268506	381	5.40
		2	1414	255340	253628	917	4.71	252977	249300	3409	3.47	255759	254816	438	5.21
		3	1342	268682	267331	863	4.30	267731	264689	2860	3.39	269393	269061	210	5.83
		4	1565	245229	243881	846	4.92	243192	237837	5179	3.64	245751	244343	367	5.24
		5	1336	277221	275980	833	4.42	277762	274254	3109	3.36	277842	277048	459	5.40
200	5	1	786	182374	181203	596	5.19	179525	177438	1331	3.83	183318	182197	485	3.50
		2	848	172119	170505	947	5.62	168021	163917	3014	3.98	172158	170861	748	3.34
		3	805	184362	182979	595	5.13	181412	178516	2361	3.83	184727	183975	403	3.50
		4	939	163832	162584	745	6.13	160146	156246	3435	4.22	164066	163083	525	3.39
		5	801	189756	188597	664	5.23	187333	185471	1179	3.82	190069	189183	334	3.54
200	10	1	393	110238	109028	507	8.87	102002	98962	1374	5.91	110528	109755	304	3.00
		2	424	102734	101595	605	9.59	92359	87400	3301	6.20	103363	102222	430	2.89
		3	402	111770	110442	537	8.82	103848	100528	2362	5.95	112273	111658	274	2.99
		4	469	95453	94544	532	10.71	85801	81481	2077	6.64	95839	95019	338	2.97
		5	400	114260	112828	586	8.82	105078	102857	1742	5.91	114585	113607	442	2.93

a – Execution time on a Pentium 4, 2.53 GHz system with 1 GB RAM

cost of mutation operator of SSGGA, mutation operator of HJ-GA, which is also the variation operator of HJ-SHC, increases with increase in K. This is due to the fact that mutation deletes more objects with increase in K, and as a result more time is spent in subsequent filling of knapsacks with unassigned objects. The cost of crossover operator of HJ-GA also increases with increase in K because crossover operator of HJ-GA first copies to the child the object assignments common to both the parents. Clearly common object assignments decreases with increase in K and as a result here also more time is spent in subsequent filling of knapsacks with unassigned objects. This explains the increase in running times of HJ-GA and HJ-SHC with increase in K. However, the cost of crossover operator of SSGGA decreases with increase in K. This is due to the fact that for small K, the first phase of crossover operator of SSGGA is more disruptive and as a result after the end of first phase, there are lesser number of objects in knapsacks, therefore more time is spent in second phase, which tries to fill the knapsacks to their capacity with unassigned objects. This decrease in crossover cost is more in comparison to increase in mutation cost with increase in K. Moreover, crossover is used 60% of times. These two factors together are responsible for the decrease in running time of SSGGA with increase in K.

4 Conclusions

In this paper we have developed a new steady-state grouping genetic algorithm (SSGGA) for the quadratic multiple knapsack problem. We have compared SSGGA with two recently proposed heuristics HJ-GA and HJ-SHC [5] on 60 QMKP instances. SSGGA outperformed HJ-GA and HJ-SHC both in terms of best as well as average solution quality. In fact, average solution values obtained by SSGGA are always better than those obtained by HJ-GA and HJ-SHC.

References

1. Garey, M.R., Johnson, D.S.: Computers and Intractability: A Guide to the Theory of NP-Completeness. W. H Freeman, San Francisco (1979)
2. Khuri, S., Bäck, T., Heitkötter, J.: The Zero/One Multiple Knapsack Problem and Genetic Algorithms. Proceedings of the 1994 ACM Symposium on Applied Computing, ACM Press, New York (1994), 188-193
3. Cotta, C., Troya, J.M.: A Hybrid Genetic Algorithm for the 0-1 Multiple Knapsack Problem. Artificial Neural Networks and Genetic algorithms 3, Springer-Verlag, Berlin (1998), 250-254
4. Yoon, Y., Kim, Y.H., Moon B.R.: An Evolutionary Lagrangian Method for the 0/1 Multiple Knapsack Problem. Proceedings of the GECCO-2005, ACM Press, New York (2005), 629-635
5. Hiley, A., Julstrom, B.A.: The Quadratic Multiple Knapsack Problem and Three Heuristic Approaches to it. Proceedings of the GECCO-2006, ACM Press, New York (2006), 547-552

6. Falkenauer, E.: New Representations and Operators for GAs Applied to Grouping Problems. Evolutionary Computation 2 (1992), 123-144
7. Falkenauer, E.: Genetic Algorithms and Grouping Problems. John Wiley & Sons, Chicester (1998)
8. Davis, L.: Handbook of Genetic Algorithms. Van Nostrand Reinhold, New York (1991)
9. Singh, A., Gupta, A.K.: Two Heuristics for the One-Dimensional Bin-Packing Problem. To appear in OR-Spectrum (2007)

A Hybrid Method for Solving Large-Scale Supply Chain Problems

Steffen Wolf and Peter Merz[*]

Distributed Algorithms Group
University of Kaiserslautern, Germany
{wolf,pmerz}@informatik.uni-kl.de

Abstract. The strategic supply chain design problem which allows capacity shifts and budget limitations can be formulated as a linear program. Since facilities are allowed to be opened or shut down during the planning horizon, this problem is in fact a mixed integer problem. Choosing the optimal set of facilities to serve the customer demands is an NP-hard combinatorial optimization problem. We present a hybrid method combining an evolutionary algorithm and LP based solvers for solving large-scale supply chain problems, which takes its power from filtering out infeasible solutions. The EA incorporating these filters is shown to be faster than the MIP solver ILOG CPLEX in most of the considered instances. For the remaining instances it finds feasible solutions much faster than the MIP solver.

1 Introduction

Consider the following optimization problem: Given a set of customers to serve, a set of products to manufacture and deliver, projections for the demand of the customers as well as for transportation and purchasing costs, the aim is to find optimal combinations of facilities to deliver the goods for a predefined number of time periods. This is a typical problem in supply chain design.

Many simple facility location problems have been studied in the past [1,2,3]. However, in real world problems the number of locations that can be opened or shut down is often implicitly limited by budget constraints. Also, in real world problems capacity limitations are usually not fixed, but can be extended.

In this work, we use a more advanced supply chain model, that introduces more freedom of choice, but also increases the complexity and therefore the optimization effort. We propose a hybrid method comprising filters and an evolutionary algorithm to approach large scale supply chain design problems. We show in experiments on various problems of different sizes how much can be gained using the filters and compare the results to ILOG CPLEX [4].

The paper is organized as follows. In Section 2, the mathematical formulation for the considered supply chain optimization problem is given. Methods for filtering out infeasible solutions and thereby reducing the search space are presented

---

[*] This work was partially supported by the Rhineland-Palatinate Cluster of Excellence 'Dependable Adaptive Systems and Mathematical Modelling'.

in Section 3. In Section 4, we present a simple evolutionary algorithm incorporating these filters and give results in Section 5. Conclusions and an outline for future research are provided in Section 6.

2   Supply Chain Optimization

Many supply chain optimization problems can be formulated as a Linear Program (LP) or a Mixed Integer Linear Program (MIP). This allows standard algorithms and also standard software to be used, e. g. CPLEX. In this work, we use the mathematical model given in [5], which is formulated as an MIP.

2.1   Problem Definition

The model uses a number of parameters, which can be best described in form of matrices. In the following, $X^t_{l,p}$ denotes the value for parameter or variable X for time period t, location l and product type p, and $X^t_{l,l',p}$ denotes the value for the arc from facility l to l'. If X does not depend on the location, the product type or the time period, the corresponding index l, p or t is not written.

Parameter $D^t_{l,p}$ contains the customer demand, $PC^t_{l,p}$ the unit purchase cost, $TC^t_{l,l',p}$ the transportation cost, $IC^t_{l,p}$ the inventory carrying cost and OC^t_l the operation cost. The shutdown cost SC^t_l, the facility setup cost FC^t_l and the cost for shifting capacity from one facility to another facility $MC^t_{l,l'}$ are limited by a budget B^t. Non-invested capital ξ^t can be saved with interest rate β^t. The capacity of a facility is limited by an upper bound $\overline{K}^t_l$ and a lower bound $\underline{K}^t_l$, the latter also denotes a minimal throughput for this facility. A unit capacity consumption factor $\alpha_{l,p}$ allows to let different product types or facilities have different influences on the capacities.

The variables in this model are the binary operational status δ^t_l of the facilities, where $\delta^t_l = 1$ denotes that facility l is operating, and $\delta^t_l = 0$ that it is not operating in time period t. Closed facilities have no capacity, and once a facility falls below the minimum capacity $\underline{K}^t_l$ it has to be closed. The other variables are the product flow $x^t_{l,l',p}$, the amount of purchased products $b^t_{l,p}$, the amount of products to be stored for the next time period $y^t_{l,p}$, and the amount of capacity to be shifted $z^t_{l,l'}$. Some values for the first time period are pre-defined, such as the amount of products held on stock, the non-invested capital and the operational status. All parameters have positive values, only the shutdown costs SC^t_l can be negative, e. g. when governmental subsidies are given.

The cost function C to be minimized is:

$$C = \sum_{\substack{t \in T \\ l \in L \\ p \in P}} PC^t_{l,p} b^t_{l,p} + \sum_{\substack{t \in T \\ l,l' \in L, l \neq l' \\ p \in P}} TC^t_{l,l',p} x^t_{l,l',p} + \sum_{\substack{t \in T \\ l \in L \\ p \in P}} IC^t_{l,p} y^t_{l,p} + \sum_{\substack{t \in T \\ l \in L}} OC^t_l \delta^t_l \quad (1)$$

Here, $T = \{1, \ldots, n\}$ is the set of time periods, P the set of product types and L the set of all facilities, e. g. plants, distribution centers and customers.

There are three pre-defined sets of facilities: $S^o \subseteq L$ the facilities to be opened, $S^c \subseteq L$ the facilities to be closed, and $L \setminus (S^o \cup S^c)$ the non-selectable facilities. The following equations give the current capacity K_l^t for all facilities:

$$\forall l \in S^c, t \in T: \ K_l^t = \overline{K}_l^1 - \sum_{\substack{\tau \in \{1,\dots,t\} \\ i \in S^o}} z_{l,i}^\tau \tag{2}$$

$$\forall l \in S^o, t \in T: \ K_l^t = \sum_{\substack{\tau \in \{1,\dots,t\} \\ i \in S^c}} z_{i,l}^\tau \tag{3}$$

$$\forall l \in L \setminus (S^o \cup S^c), t \in T: \ K_l^t = \overline{K}_l^t \tag{4}$$

The capacity for facilities in S^c is defined by the parameter $\overline{K}_l^1$ and can only be reduced by shifting capacity to another facility (2). New facilities start with no capacity and have to receive capacity shifts before they can operate (3). The capacities of the remaining facilities is given in the parameter $\overline{K}_l^t$ and cannot be changed (4). Using these equations the remaining constraints can be formulated as:

$$\forall l \in L, p \in P, t \in T: b_{l,p}^t + \sum_{l' \in L \setminus \{l\}} x_{l',l,p}^t + y_{l,p}^{t-1} = D_{l,p}^t + \sum_{l' \in L \setminus \{l\}} x_{l,l',p}^t + y_{l,p}^t \tag{5}$$

$$\forall l \in L, t \in T: \underline{K}_l^t \delta_l^t \leq \sum_{p \in P} \alpha_{l,p} \left(b_{l,p}^t + \sum_{l' \in L \setminus \{l\}} x_{l',l,p}^t + y_{l,p}^{t-1} \right) \leq K_l^t \leq \overline{K}_l^t \delta_l^t \tag{6}$$

$$\forall l \in S^c, t \in T: K_l^t \geq \delta_l^t \epsilon \tag{7}$$

$$\forall t \in T: \sum_{\substack{l \in S^c \\ l' \in S^o}} MC_{l,l'}^t z_{l,l'}^t + \sum_{l \in S^c} SC_l^t \left(\delta_l^{t-1} - \delta_l^t \right) + \sum_{l \in S^o} FC_l^t \left(\delta_l^{t+1} - \delta_l^t \right) + \xi^t$$

$$= B^t + (1 + \beta^{t-1}) \cdot \xi^{t-1} \tag{8}$$

$$\forall t \in T, l \in S^c: \ \delta_l^t \geq \delta_l^{t+1} \tag{9}$$

$$\forall t \in T, l \in S^o: \ \delta_l^t \leq \delta_l^{t+1} \tag{10}$$

In short, constraints (5) state that all demands have to be fulfilled either by purchasing products, delivery from another location or by having them stored in the previous time period. Constraints (6) ensure that a facility cannot produce, receive or store more than its capacity in each time period. Moreover, its maximum and minimum allowed capacity cannot be exceeded. Constraints (7) (with $\epsilon > 0$ sufficiently small) ensure that a facility has to be closed when all its capacity has been removed. Constraints (8) give the budget limitations, which include shutdown and setup costs as well as moving capacity costs, and allow unused budget to be saved (with interest) for a later time period. Finally, constraints (9) and (10) state that facilities that have been closed cannot be re-opened, and facilities that have been opened cannot be closed again. Obvious constraints, such that all values have to be real and positive, have been left out.

The problem can be reduced to a static uncapacitated facility location problem which was shown to be NP-hard in Cornuéjols *et al.* [6].

2.2 Related Work

In [5] various small problems for this model and a slightly improved version were solved to optimality by off the shelf commercial software. Velásquez and Melo [7] have introduced variable neighborhood search heuristics to approach larger problems. Here, a linear programming (LP) solver was used to solve the LP subproblem that is obtained by fixing all δ_l^t to values determined by the heuristics. Based on the total cost, the heuristics would then slightly change the δ-values and calculate the new δ-combination. However, no explicit effort has been made to filter out infeasible combinations. So, the LP solver often takes a non-negligible time to report the infeasibility of the considered combination. Since an infeasible combination is undesired, this though small calculation time can be considered as wasted.

3   Search Space Reduction

We propose a way of improving the heuristics in [7]. Again we separate the generation of δ-combinations from the LP calculations. The generation of combinations can be seen as a combinatorial optimization problem with a very complex cost function. Since most computation time is used in the LP calculations we seek to avoid this calculation as often as possible.

Our method of choice is filtering. Once a δ-combination has been chosen, a number of simplified constraint checks are applied. Only combinations passing these checks will be presented to the LP solver. This way, we can avoid many unnecessary calls to the LP solver, allowing it to reuse information from previous runs without the interruption caused by an infeasible combination. Also, by using simplified and explicit constraints, these checks can be carried out faster.

3.1 Shutdown and Setup Costs

A first check that can be applied is to calculate the sum of shutdown and setup costs and check whether the amount stays within the budget limits:

$$\forall t \in T : \sum_{l \in S^c} SC_l^t \left(\delta_l^{t-1} - \delta_l^t \right) + \sum_{l \in S^o} FC_l^t \left(\delta_l^{t+1} - \delta_l^t \right) \leq B^t + (1 + \beta^{t-1}) \cdot \hat{\xi}^{t-1}$$

Here, $\hat{\xi}$ is the remaining capital when capacity shift costs are ignored. This inequality is a relaxation of (8). The check filters out all combinations where too many changes are scheduled. In real world examples only a very small number of facilities are supposed to be opened, but a larger number of possible locations may be provided to choose from. Once a combination is chosen this check can be applied iteratively for each time period. When the accumulated costs exceed the budget (plus accumulated interests) the combination can be discarded, aborting all following checks. Also, a specific facility can be identified whose opening or shutdown caused the check to fail.

3.2 Capacity Shifts

In the model, capacity can only be shifted from existing to new facilities. Also, all capacity from a facility that is to be closed has to be shifted to new facilities. Since a maximum capacity is given, this may not always be possible. A quick check can reveal whether there are enough new facilities to receive the capacity of the existing facilities. Again, this has to be verified for each time period.

This analysis also gives an amount of capacity $\hat{z}$ which has to be shifted, thus creating capacity shift costs that have to be covered by the budget. Since these costs depend on time period and involved facilities, we cannot determine the exact cost, but only give lower bounds. A first lower bound takes the minimum of all entries in MC. A better lower bound takes account of the facilities that need to lose or receive capacity and uses the minimum of only those entries in MC. A more sophisticated lower bound may also include the time period. However, capacity shifts can be scheduled in any time before closing the facility, so this does not not yield much. Denoting this minimum as $MC_{\min}$, these checks can be formulated as:

$$MC_{\min} \cdot \hat{z} + \sum_{l \in S^c} SC_l^t \left(\delta_l^{t-1} - \delta_l^t \right) + \sum_{l \in S^o} FC_l^t \left(\delta_l^{t+1} - \delta_l^t \right) \leq B^t + (1 + \beta^{t-1}) \cdot \hat{\xi}^{t-1}$$

3.3 Customer Demands

The parameter D gives the customer demands. All demands have to be fulfilled. However, customers may buy the needed products from an external supplier. This is accounted for by introducing some form of penalty costs. In order for the customer demands to be fulfilled there has to be an operating facility and a path from this facility to the customer. If this path passes other facilities (e. g. plants – warehouses – customers), they also have to be operating. Calculating all paths for all customers may be expensive, considering that the path may reach over multiple time periods since products can be stored. We therefore determine for each customer and product type only the set of facilities that can deliver the goods directly. Out of this set at least one facility has to be operating.

4 Evolutionary Algorithm for the Supply Chain Problem

In this section we present a simple Evolutionary Algorithm (EA) that incorporates these checks. The algorithm is shown in Fig. 1. In this simple EA we use only mutation. The population size is fixed at μ. In each generation a mutation operation is used to generate λ valid children. After the mutation operator is applied

```
function EA( κ, λ, μ: Integer );
begin
   for i := 0 to μ−1 do p[i] := Init ();
   for iter := 0 to κ−1 do
   begin
     for i := 0 to λ−1 do
     begin
       repeat
         p[μ+i] := Mutate( p[i mod μ] );
         if CheckFilters(p[μ+i]) then LPsolver(p[μ+i]);
       until isFeasible (p[μ+i]);
     end;
     p := Select (p);
   end;
   return Best(p);
end;
```

Fig. 1. The Evolutionary Algorithm Framework

the resulting combination is presented to the filter. If one of the checks fails, the combination is removed and another offspring is created. Only those combinations that pass all checks are given to the LP solver. The combination is also discarded, if the LP solver states infeasibility. Once λ valid offspring combinations are found and evaluated, the combinations to form the next generation are selected. Out of the μ original and the λ offspring combinations, the μ best individuals are chosen, following a $(\mu + \lambda)$ selection paradigm. This procedure is repeated κ times.

To build the initial generation, random combinations can be used as well as combinations based on the LP-relaxation solution of the supply chain problem. In the random approach, a combination is created by scheduling an opening/ closing of a facility with probability p, choosing the time period at random. When no information about the maximum number of possible shutdowns and openings is available, this procedure has to be repeated for some possible values of p. Once a feasible combination is found, p can be adjusted to match the best known combination. The LP-relaxation approach works by solving the LP problem which results when dropping the constraint that all δ have to be binary. The solution found may still assign binary values to some δ. The remaining non-binary values can be used to 'guess' the values of other δ. A binary δ-combination can be generated from this knowledge by random assignment of 0 or 1, using the δ-values as a probability. However, these combinations tend to be infeasible since they often schedule more openings/shutdowns than the budget allows. Reducing or limiting the number of opening/shutdowns helps to find feasible combinations.

Instead of storing all δ-values, we use a more compact encoding. For each selectable facility l we store the time period γ_l of its opening/shutdown. Thus, each individuum can be described by only $s =| S^o \cup S^c |$ integer variables instead of $| T | \cdot | L |$ binary variables. The operational status δ can easily be extracted from this encoding using the following formula:

$$\delta_l^t = 1 \leftrightarrow (l \in S^o \wedge t \geq \gamma_l) \vee (l \in S^c \wedge t < \gamma_l) \vee l \in L \setminus (S^o \cup S^c) \tag{11}$$

The mutation operator used in this EA changes the time period for the opening/shutdown of a random facility to a random value $\gamma_l \in [2, n+1]$. To schedule an opening/shutdown for period $\gamma_l = n+1$ denotes that the facility should not be opened/closed during the planning horizon. We have also tried other mutation operators, but the multiple consecutive application of the described operator gave the best results.

5   Experiments

To show the effects of these filters we ran experiments on various supply chain instances. Since standard MIP software like ILOG CPLEX [4] can solve smaller problems in very short time, we concentrated on larger instances. Finding real world data for those larger instances proved to be difficult, so we used randomly generated instances as described in [8]. Nine instances were generated with up to 70 selectable and 55 non-selectable locations, 5 product types and 8 time periods. In all instances there are two kinds of distribution centers (DCs). All products have to be transported from the plants to central DCs, and from there to regional DCs before they arrive at the customer sites. Each DC can store products for later time periods. The DCs are the selectable facilities in these instances, all other facilities (such as customers and plants) are non-selectable. The capacities of the locations are limited, but those limits can be changed during the planning horizon according to the model.

In a first set of experiments we were interested in the number of δ-combinations that remain when using the filters. Table 1 shows the results for six small instances (H1, ..., H6). Since this analysis is very expensive – it requires calculating or checking all δ-combinations – it was not applied to the larger instances. The first line gives the number of possibilities to schedule openings/shutdowns for $s = |\, S^o \cup S^c\,|$ locations in $n = |\, T\,|$ time periods: n^s. Most of these combinations are infeasible due to the budget constraints. After the first check the number of combinations is already drastically reduced. Another large amount of combinations is filtered out by the capacity shift checks. The remaining number is shown in the third line. However, as stated in Section 3, the infeasibility of some combinations cannot be recognized by these simple checks. The number of feasible combinations is shown in the fourth line. This number was determined by handing all remaining combinations to the LP solver. The last line shows the influence of the demand satisfiability check. This check filters out combinations with high penalty costs regardless of their feasibility, so the remaining number of combinations is often smaller than the actual number of feasible combinations. Only in problem H3 no combination lead to unfulfilled demands.

The effect of unfulfilled demands can be seen in Fig. 2. The figure shows the calculated costs for some random combinations for problems H1 and N7. In H1 many combinations can be found with costs around $C \approx 10^6$, near the optimum (953 824), but there are some "bands" of combinations with higher costs. All

Table 1. Number of δ-combinations remaining after the filters

Problem	H1	H2	H3	H4	H5	H6
Total	3^{30}	4^{30}	3^{30}	3^{45}	3^{30}	3^{30}
Shutdown/Opening Cost Check	278 053	8 881 793	101 626	1 395 442	106 245	158 415
Capacity Shift Check	33 938	517 723	11 747	161 352	9 410	11 890
Feasible Combinations	25 539	322 352	8 321	111 009	6 795	7 671
Demand Satisfiability Check	21 052	299 508	11 747	134 802	3 997	8 072

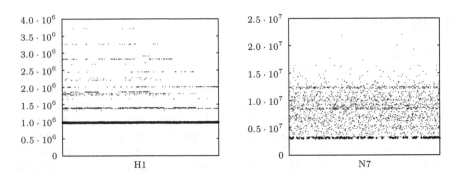

Fig. 2. The distribution of the objective function values for problems H1 and N7

these combinations have unfulfilled demands, because some required facility was closed or not opened, which generates penalty costs. In problem N7 the bands are not that clear, but they also exist here. Problem N7 shows even more bands than in H1. In fact, every dot in this plot above $0.35 \cdot 10^7$ belongs to a higher band. We have discovered these bands in all considered problems, with the exception of H3. An explanation for these bands is the possibility to close (or not open) required facilities. The longer a required facility is closed the higher is the penalty cost. The penalty cost is supposed to be much larger than the differences between the feasible solutions. So for each time period and each required facility a different penalty is applied, which creates the observed bands. The demand satisfiability check filters out these expensive combinations, leaving only the lowest band, marked by bold dots in the figure. These remaining combinations are only up to 10 % more expensive than the optimal or – in case of the larger problems – the best known solution.

In the next set of experiments we were interested in the overall performance of our proposed EA. We fixed $\mu = 5$ and $\lambda = 35$, and used CPLEX 10.1 to evaluate the fitness of each δ-combination, i.e. solve the underlying LP. The number of generations was set to $\kappa = 50$, so CPLEX had to evaluate at least $\mu + \lambda \cdot \kappa = 1755$ combinations. These parameters are a compromise between computation time and solution quality. All CPU times reported here refer to a 3 GHz Pentium IV. The results are averaged over 10 runs.

Figure 3 shows the averaged behavior of the EA for two different problems. In each case the first feasible solution is found almost instantly. In problem N7

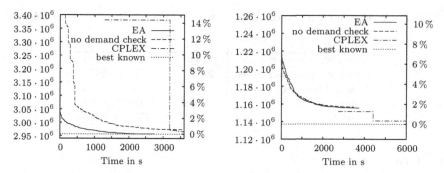

Fig. 3. Behavior of the EA with and without demand check for problems N7 and N19

the first solution already exceeds the best known solution by only 3 %. This best known solution is reached in 4 of the 10 EA runs after 30-40 min, and also by CPLEX after 25 h. The figure also shows the behavior of the EA when the demand check is not active. Without this check the cost of the first solution is almost four times as high as the best known solution, also the improvement is very slow. To compare the EA to standard software we show the behavior of CPLEX in the same plot. CPLEX in its standard configuration does not find a feasible solution for about 20 min, and the first solution it finds is worse than the first solution of the EA.

In problem N19 the influence of the demand check is negligible. Here, only a small amount of δ-combinations can be found in higher bands, and therefore be cut off. The EA finds a good starting solution and steadily improves it. However, it does not reach the best known solution, but converges between 1.5 % and 1.8 % above it. CPLEX takes about 45 min to find a feasible solution, but this time it is already better than the best solution found by the EA. The best known solution was found by CPLEX after four days.

Problem instance N19 was the worst instance for the EA. Problem N20 showed similar results like N7. We therefore omit the plot. The main difference to N7 is that the computation times are roughly 5 times as high, due to the problem size. However, the best known solution in N20 was found in an EA run after about 4 h, while CPLEX takes about 3 h to find a first feasible solution and does not find the best known solution in the first 30 h.

Table 2 shows the average number of combinations that were filtered out by the proposed filters in an EA run. These numbers are significantly lower than in Table 1, but the filters are still useful, since they filter out about 100 times more combinations than feasible ones, leaving only a small percentage of unnecessary calculations. In our experiments an LP evaluation took about 1-8 s, depending on the problem. An LP run on an infeasible combination took only 0.1-0.8 s, so it is roughly ten times faster. However, CPLEX cannot start from a good LP solution in the next evaluation, so the evaluations after such infeasible runs are slowed down. The advantage of the filters is clear. They can be applied thousand times a second and do not affect the underlying LP solver. Without these filters the EA would show very poor results.

Table 2. Average number of δ-combinations filtered out by the filters in an EA run

Problem	N7	N19	N20
Shutdown costs too high	191068	193337	148843
Capacity shifts impossible	22108	26967	19450
Demand not met	1622	18	2
Still infeasible	32	16	94
Feasible	1755	1755	1755

6   Conclusions

We have proposed an EA using filters to solve large-scale supply chain design problems. We have showed that the EA can find feasible solutions very quickly. The filters reduce the search space significantly and allow the EA to even find the optimal solution in some problems. However, in problems that show many combinations in the lower band, the demand filter cannot be applied efficiently, so the EA may not find the optimal solution and is outperformed by CPLEX. Our method still has an advantage in these cases, since it finds feasible solutions and even good solutions long before CPLEX. These can be given to CPLEX as a starting solution, in order to improve the performance of the MIP solver. This is an issue of future research.

References

1. Nemhauser, G.L., Wolsey, L.A.: Integer and Combinatorial Optimization. John Wiley & Sons, New York (1988)
2. Resende, M.G.C., Werneck, R.F.: A hybrid multistart heuristic for the uncapacitated facility location problem. European Journal of Operational Research **174** (2006) 54–68
3. Zhang, J.: Approximating the two-level facility location problem via a quasi-greedy approach. In: SODA'04: Proceedings of the 15th annual ACM-SIAM symposium on Discrete algorithms, Philadelphia, PA, USA, Society for Industrial and Applied Mathematics (2004) 808–817
4. ILOG S.A.: ILOG CPLEX User's Manual (2006) Gentilli, France. http://www.cplex.com/
5. Melo, M.T., Nickel, S., Saldanha da Gama, F.: Dynamic multi-commodity capacitated facility location: a mathematical modeling framework for strategic supply chain planning. Computers & Operations Research **33** (2006) 181–208
6. Cornuéjols, G.P., Nemhauser, G.L., Wolsey, L.A.: The uncapacitated facility location problem. In Mirchandani, P.B., Francis, R.L., eds.: Discrete Location Theory. Wiley, New York (1990) 119–171
7. Velásquez, R., Melo, M.T.: Solving a large-scale dynamic facility location problem with variable neighbourhood and token ring search. In: Proceedings of the 39th ORSNZ Conference, Auckland, NZ (2004)
8. Melo, M.T., Nickel, S., Saldanha da Gama, F.: Large-scale models for dynamic multi-commodity capacitated facility location. Technical Report 58, Fraunhofer Institut for Industrial Mathematics (ITWM), Kaiserslautern, Germany (2003) Available at http://www.itwm.fhg.de/

Crossover Operators for the Car Sequencing Problem

Arnaud Zinflou, Caroline Gagné, and Marc Gravel

Université du Québec à Chicoutimi, 555 boulevard de l'université, Chicoutimi,
Qc, G7H2B1, Canada
{arnaud_zinflou, caroline_gagne, marc_gravel}@uqac.ca

Abstract. The car sequencing problem involves scheduling cars along an assembly line while satisfying as many assembly line requirements as possible. The car sequencing problem is NP-hard and is applied in industry as shown by the 2005 ROADEF Challenge. In this paper, we introduce three new crossover operators for solving this problem efficiently using a genetic algorithm. A computational experiment compares these three operators on standard car sequencing benchmark problems. The best operator is then compared with state of the art approach for this problem. The results show that the proposed operator consistently produces competitive solutions for most instances.

1 Introduction

The car sequencing problem became important in the production process of most car manufacturers when mass customization replaced mass standardisation. The production line of a modern car factory can be viewed as a linear manufacturing process generally composed of three consecutive workshops: the *body fabrication shop*, the *paint shop* and the *assembly shop*. In the literature we find a « standard » version of the problem which deals only with assembly shop requirements. In this workshop, each car is characterized by a set of different *options O* (sunroof, ABS, air-conditioning, etc.) among which some may require more work [1]. To ensure smooth operations in the assembly shop, cars requiring high work-content must be distributed throughout the assembly line. This requirement may be formalized by r_o/s_o ratio constraints that state that any subsequence of s_o cars must include at most r_o cars with the option o. When a ratio constraint is exceeded in a subsequence, there is a *violation*. In order to simplify solution, cars requiring the same configuration of options are clustered into the same *car class*. For each of the V classes thus created, we know exactly the number of cars to produce. These quantities engender production constraints which state that exactly c_v cars of the v class must be produced. Then, the car sequencing problem involves finding the order in which nc cars from different classes should be produced in order to minimize violations. This problem has been shown to be NP-hard in the strong sense [2]. A detailed description of the formulation of both the industrial and standard version of the car sequencing problem can be found in [3, 4].

The standard car sequencing problem has been widely studied since its first introduction in the middle 80's and comprehensive surveys on the problem and the methods used to solve it can be found in literature [5, 6]. Most recent works have focused on neighbourhood search [7, 8] and on various ant colony optimization algorithms

C. Cotta and J. van Hemert (Eds.): EvoCOP 2007, LNCS 4446, pp. 229–239, 2007.
© Springer-Verlag Berlin Heidelberg 2007

[4, 5, 9-10]. One notes that few authors have proposed genetic algorithms, save for Warwick and Tsang [11] and most recently Terada *and al.* [12]. This situation may be explained by the difficulty of defining specific and efficient genetic operators for the problem. In fact, traditional genetic operators are generally defined for traveling salesman problems (TSP), binary representation problems [13, 14] or real codification problems [15] and can not deal adequately with the specificities of car sequencing.

In this paper, we introduce three new crossover operators to efficiently solve the car sequencing problem with a genetic algorithm. The remainder of the paper is organized as follows: the next section briefly presents the genetic algorithm and its application to car sequencing; in Section 3, we describe the three new crossover operators proposed; in Section 4, we present the computational experiment on CSPLib's benchmarks. Some concluding remarks are drawn in the final section.

2   Genetic Algorithms (GA)

Genetic algorithms are stochastic algorithms based upon the natural selection theory of C. Darwin and the genetic inheritance laws of G. Mendel. The basic concepts of genetic algorithms were first presented by Holland [16] for mathematical optimization and popularised thereafter by Goldberg [17]. The application field of GA techniques is wide, and they are particularly successful in solving many hard combinatorial optimization problems [15, 17-22].

To our knowledge, Warwick and Tsang [11] were the first to apply genetic algorithms in solving the car sequencing problem. In their approach, at each generation, selected sequences are combined using a *uniform adaptive crossover* (UAX); as the created offspring may not satisfy the production constraints, they are greedily repaired; after repair, each offspring is hill-climbed by a standard swap function. In recent work, Terada [12] proposed a classical genetic algorithm for solving the car sequencing problems where recombination of two individuals is performed by one-point crossover and explored the possibility of combining it with *Squeaky-Wheel Optimization* (SWO) techniques.

3   Three New Crossover Operators

In order to present the different crossover operators, we must define two important concepts used by these operators: the *difficulty of a class* and the *interest* to add a car of class v at the position i in the sequence.

The difficulty D_v of a class v is obtained by summing the *utilization rates* of the options (utr_o) required by the class:

$$D_v = \sum_{k=1}^{|o|} o_{vk} \; utr_o \; . \tag{1}$$

where $o_{vk} = \{0,1\}$ indicates if the cars of class v require the option k. Formally, the utilization rate of an option o can be expressed as a ratio between the number of cars requiring this option (nb_o) and the maximum number of cars that can receive this option so that the r_o/s_o is satisfied, i.e:

$$utr_o = {nb_o} \Big/ {\lceil nc \cdot r_o / s_o \rceil} \cdot \tag{2}$$

A utilization rate greater than 1 indicates that the capacity of the station will inevitably be exceeded. On the other hand, a rate near 0 indicates that the demand is very low with respect to the capacity of the station. However, even if all utilization rates are less than or equal to 1, a feasible solution does not necessarily exist. Note that the utilization rate for each option is computed dynamically as proposed by Gottlieb *et al.*[6].

The interest I_{vi} to add a car of class v at the position i given the cars already assigned in the sequence is then given by:

$$I_{vi} = \begin{cases} D_v & \text{if } NbNewViolations_{vi} = 0 \\ -NbNewViolations_{vi} & \text{otherwise} \end{cases} \tag{3}$$

where $NbNewViolations_{vi}$ indicates the number of new violations caused by the addition of a car of v at the position i in the sequence.

3.1 Interest Based Crossover (IBX)

The first crossover operator proposed is inspired by the PMX operator [23]. Fig. 1 illustrates our approach using a small example. The first step of the IBX operator is to randomly select two cut points in both parents P_1 and P_2. The substring 351 between the two cut points in P_1 is then directly pasted in the same position in the offspring. Then, two non order lists (L_1 and L_2) are constituted using the substrings $\{3, 2\}$ and $\{4, 5, 6\}$ of P_2. However, one effect of this process is that the production requirements will not always be satisfied. In our example, one notes that production constraints for class 2, 3, 4 and 5 are no longer satisfied. To correct this, a replacement of the class 3 and 5, of which the number exceeds the production constraint, by the class 4 and 2, of which the number is now less than the production constraint, is randomly applied in the lists L_1 and L_2 at the second step. Finally, the last step rebuilds the beginning and the end of the offspring using the two lists. To do that, the class of cars $\in L_1$ are ordered using their interests from the first cutting point to the beginning of the sequence. Therefore, the class v to place now is given by:

$$v = \begin{cases} \arg \max \{I_{vi}\} & \text{if } p \leq 0.95 \\ V & \text{otherwise} \end{cases} \tag{4}$$

where p is a random number between 0 and 1 and V is chosen in a probabilistic manner. To determine V, the roulette wheel principle is used [17] within the class for which the addition caused the fewest new violations.

The same process is used to order the cars from the second cutting point to the end of the sequence using L_2. A second offspring is generated by simply inverting the roles of the parents.

This crossover technique, contrary to PMX, does not try to preserve the absolute position of the cars when they are copied from the parents to the offspring. IBX tries

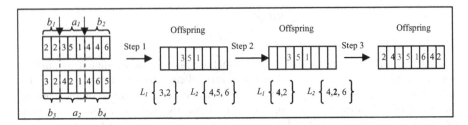

Fig. 1. Schematic of IBX crossover

rather to keep the cars in the same area of the chromosome as the one they occupied in one of the two parents. In fact, the number of classes that do not inherit their assigned area from one of the two parents is at most equal to the length of the string between the two cut points.

3.2 Uniform Interest Crossover (UIX)

The second crossover operator is a variant of the uniform crossover proposed by Syswerda [24]. The first step of the UIX approach is achieved by creating a random bit crossover mask of the same length as the parents. The bits valued at 0 in the crossover mask indicate the classes of cars which are taken from parent 1 and pasted at the same position in the offspring. In Fig. 2, the positions {2, 3 and 7} of the offspring are filled with class 2, 1 and 4 of P_1. The others classes (3, 4, 2, 4 and 5) inherited from parent 2 are then used to constitute a non-order list L. Once again, the production requirements will not always be satisfied after this step. To ensure that the production requirement will be satisfied in the offspring, the next step consists in finding, using L and the class already included in the offspring, which ones exceed or are less than the production constraint. Hence in our example, class 4 exceeds the production constraint while class 6 is less than this production constraint. Then, a random replacement is applied in the list or in subset of classes already included in the offspring, in order to eliminate exceeding classes and to satisfy the production constraints. In Fig. 2, a class 6 replaces class 4 in the offspring. Finally, the last step of the crossover consists of filling the remaining positions of the offspring with classes from L. In this step, a class v is chosen to fill a position i according to its interest I_{vi} as described in IBX crossover.

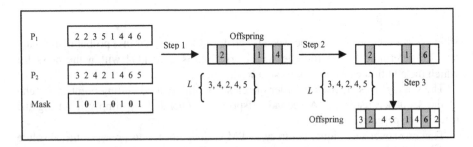

Fig. 2. Schematic of the UIX crossover

A second offspring is generated using the same approach by inverting the roles of the parents. The particularity of this technique is that class of cars inherited from the second parent can be reorganized in the offspring chromosome.

3.3 Non Conflict Position Crossover (NCPX)

The last crossover operator attempts to use non conflict positions of the parents to generate the offspring. To this end, a random number nb_g is chosen between 0 and *nbpossconflict* where *nbpossconflict* indicates the number of non conflict positions found in parent 1 (P_1). The number nb_g is used to indicate how many "good" classes of car will inherit their positions from P_1 in the offspring. A random starting point (Pos_d) is selected between the beginning and the end of the offspring. The class located in non conflict positions are then copied from the first parent to the offspring starting at Pos_d to the end of the chromosome. If the number of classes included in the offspring is less than nb_g, the copy process restarts, this time starting at the beginning of the offspring chromosome to Pos_{d-1}. Note that in all cases, the copy process is stopped as soon as nb_g cars are copied. The remainder of the classes from P_1 are then used to constitute the remaining car class list L. Thereafter, another random position (Pos) from which the remaining position of the offspring chromosome will be filled is chosen. Finally, the classes in L are assigned to the offspring according to Equation (4). However, one notes that in case of ties on Arg max$\{I_{vi}\}$, if one class in the tie occupies the current position in P_2 without conflicts, this class is chosen to be inherited in the offspring. If no class of cars can be inherited from P_2, ties are broken randomly. The operation of the NCPX operator is illustrated in Fig. 3 for two individuals $P_1 =$ 21352446 and $P_2 = 32621454$. Let us assume that there are 5 positions without conflicts on P_1 and that the number nb_g and Pos_d are respectively equal to 4 and 3. Accordingly, the class of cars 5, 4, 4 and 2 are copied to the offspring. The remaining classes 2, 3, 1 and 6 from P_1 are used to constitute the initial list L. Finally, if we assume that $Pos = 7$, we can fill the remaining positions in the offspring respectively with classes 3, 2, 6 and 1 of L starting at Pos. Hence, positions 1, 2 and 6 are directly inherited from P_2. The final offspring generated from P_1 and P_2 is then $E_1 = 22651443$.

As for the two other crossovers, a second offspring is generated by simply inverting the roles of the parents. The objective of the NCPX operator is to emphasize non conflicts positions from the parents in order to minimise the number of relocated cars.

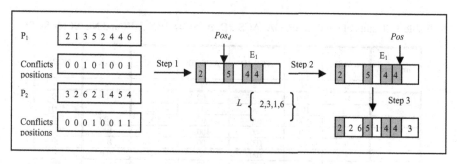

Fig. 3. Schematic of the NCPX crossover

4 Computational Experiments

The performance of the proposed crossover will be illustrated using three test suites of car sequencing problem available in the benchmark library CSPLib (http://www.csplib.org/). These instances have between 100 and 400 cars. The first set (SET1) contains 70 problems of 200 cars having 5 options and from 17 to 30 classes. These 70 instances are divided into 7 groups per utilization rate and for which there is at least one feasible solution. The second set (SET2) [25] is composed of 9 harder instances some of them are feasible, whereas some others are not. These instances have 100 cars to sequence, 5 options and from 18 to 24 classes. Finally, the last set (SET3) proposed by Gravel *and al.*[4] contains 30 difficult instances from 200 to 400 cars with the same characteristics as those of the SET2. The algorithms proposed here are all implemented in C^{++} and compiled using Visual Studio .Net 2005. The computational test for the three crossovers ran on a Pentium Xeon (3.6 Ghz and 1 Gb of RAM) with Windows XP.

For all test problems, parameters N, p_c, p_m, NbGen indicating respectively the population size, the crossover probability, the mutation probability and the maximum number of generations of our genetic algorithm have been assigned to the following values: 250, 0.8, 0.09 and 700. These values were chosen empirically in order to obtain fair comparison with the other algorithms in competition. It is also important to note that for the mutation four basic transformations operators are used here: *swap, reflection, shuffle and displacement.*

Table 1 reports the results of the three proposed approaches for the 70 instances of SET1 and compares them to those of 2 other genetic algorithms (GAcSP [11] and GA[12]) and to those of two ant colony optimization heuristics (ACS-2D [4] and ACS-3D [9]) from the literature. In this table, we present the name of the group instance and the percentage of successful runs for each algorithm. Note that GAcSp have not been tested for all the 70 instances. It is therefore hard to compare our approach directly with these two others genetic algorithms. We can however look at the conclusions of the author's experiments; they show that even if these two methods solve instances with small utilization rates the number of successful runs is severely decreased with higher utilization rates. We observe that all these instances are trivially solved by both ACS algorithm and by the three crossovers proposed.

Table 2 reports results of ACS-2D, ACS-3D, IBX, UIX and NCPX for SET2. Each instance was solved 100 times by each algorithm. The table presents the name of the

Table 1. Results of the GAcSP, GA, ACS-2D, ACS-3D, IBX, UIX and NCPX for SET1

Instance	GAcSP	GA	ACS-2D	ACS-3D	IBX	UIX	NCPX
	%	%	%	%	%	%	%
60-*	19	100	100	100	100	100	100
65-*	-	100	100	100	100	100	100
70-*	23	100	100	100	100	100	100
75-*	-	80	100	100	100	100	100
80-*	9	16	100	100	100	100	100
85-*	-	2	100	100	100	100	100
90-*	-	1	100	100	100	100	100

Table 2. Results of the ACS-2D, ACS-3D, IBX, UIX and NCPX for SET2

Instance	Best known solution	ACS-2D Mean violations	ACS-3D Mean violations	IBX Mean violations	UIX Mean violations	NCPX Mean violations
10_93	3	4.24	4.03	4.00	3.99	3.55
16_81	0	0.12	0.58	0.65	0.77	0.03
19_71	2	2.08	2.04	2.26	2.75	2.00
21_90	2	2.63	2.02	2.25	2.47	2.00
26_82	0	0.00	0.00	0.01	0.31	0.00
36_92	2	2.29	2.03	2.41	2.61	2.00
4_72	0	0.00	0.01	0.00	0.00	0.00
41_66	0	0.00	0.00	0.00	0.00	0.00
6_76	6	6.00	6.00	6.00	6.00	6.00

Table 3. Results of the ACS-2D, ACS-3D, IBX, UIX and NCPX for SET3

Instance	Best known solution	ACS-2D Mean violations	ACS-3D Mean violations	IBX Mean violations	UIX Mean violations	NCPX Mean violations
200_01	0	3.80	2.00	3.13	5.36	1.23
200_02	2	4.14	2.38	3.93	5.91	2.94
200_03	4	8.90	7.45	11.47	13.47	7.41
200_04	7	9.86	7.87	7.47	10.85	7.39
200_05	6	8.81	7.29	7.71	8.87	6.69
200_06	6	6.87	6.03	6.00	9.50	6.00
200_07	0	2.99	0.67	3.44	3.74	0.15
200_08	8	8.00	8.00	8.00	8.00	8.00
200_09	10	11.85	10.97	11.40	13.13	10.53
200_10	19	21.44	20.19	20.72	24.28	21.40
300_01	0	5.33	3.89	5.55	5.67	2.79
300_02	12	13.15	12.57	14.78	15.10	12.02
300_03	13	14.54	13.85	15.92	16.73	13.11
300_04	7	10.33	8.69	10.98	12.03	7.71
300_05	29	40.55	42.54	39.39	43.42	42.83
300_06	2	7.59	5.79	8.24	8.31	5.30
300_07	0	2.89	0.97	3.78	3.92	0.08
300_08	8	9.17	8.95	9.11	10.00	8.00
300_09	7	9.05	8.00	9.40	10.80	7.36
300_10	21	34.63	32.56	32.23	32.33	28.48
400_01	1	3.01	3.50	2.98	4.50	1.81
400_02	15	23.28	23.82	23.41	23.52	19.31
400_03	9	11.65	13.64	12.21	12.25	10.79
400_04	19	21.96	20.38	20.37	29.26	19.12
400_05	0	3.48	2.68	5.06	5.15	0.00
400_06	0	4.20	1.53	4.44	6.79	0.16
400_07	4	7.65	8.68	5.59	6.41	4.72
400_08	4	11.54	12.67	7.22	7.40	4.73
400_09	5	17.98	16.01	17.38	17.45	10.58
400_10	0	4.24	2.66	4.79	6.09	0.71

instance, the best known solution and the average number of violations found. By comparing the three crossover operators, one observes that NCPX crossover outperforms the two others for 6 of the 9 instances and obtains similar results on the 3 remaining instances. The efficiency of NCPX crossover is confirmed by comparing its results to those of the two ACS algorithms. Indeed, GA with NCPX crossover outperform the two ACS on 5 of the 9 instances (10_93, 16_81, 19_71, 21_90 and 36_92) while obtaining exactly the same results on the 4 other instances.

Table 3 summarizes the results obtained by the crossover operators and the 2 ACS algorithms for SET3 in the same way as Table 2. Once again, each instance was solved 100 times by each algorithm. One notes that both IBX and NCPX crossover outperform UIX on all problems except for instance 200_08 for which the results are equal. In comparing IBX to NCPX, we observe that the results for NCPX are better than those for IBX on 26 of the 30 instances, equal on 2 (200_06 and 200_08) and worse only on instances 200_10 and 300_05. Once again the efficiency of the NCPX approach is confirmed by comparing its results to those of the 2 ACS algorithms. Indeed, one notes that GA with NCPX outperforms the ACS-2D algorithm on 28 instances, obtains equal results for instance 200_08, and is worse for 300_05. Comparing ACS-3D to GA with NCPX, we observe that the GA obtains better results for 26 instances, is equal for instance 200_08 and is worse for 3 instances (200_02, 300_05, 200_10). We note, by the way, that the gap between the algorithms increases with size of the problems.

Table 4. Results of the ACS-2D, ACS-3D and NCPX all with local search for SET2 and SET3

Instance	Best known solution	ACS-2D + local search	ACS-3D + local search	NCPX+local search
		Mean conflicts	Mean conflicts	Mean conflicts
10_93	3	3.80	3.37	3.09
16_81	0	0.00	0.03	0.00
19_71	2	2.00	2.00	2.00
21_90	2	2.00	2.00	2.00
26_82	0	0.00	0.00	0.00
36_92	2	2.00	2.00	2.00
4_72	0	0.00	0.00	0.00
41_66	0	0.00	0.00	0.00
6_76	6	6.00	6.00	6.00
200_01	0	1.00	0.41	0.21
200_02	2	2.41	2.00	2.01
200_03	4	6.04	5.76	5.28
200_04	7	7.57	7.00	7.00
200_05	6	6.40	6.16	6.00
200_06	6	6.00	6.00	6.00
200_07	0	0.00	0.00	0.00
200_08	8	8.00	8.00	8.00
200_09	10	10.00	10.00	10.00
200_10	19	19.09	19.02	19.06
300_01	0	2.15	1.87	0.11
300_02	12	12.02	12.01	12.00
300_03	13	13.06	13.02	13.00
300_04	7	8.16	7.70	7.32
300_05	29	32.28	32.53	32.23
300_06	2	4.38	3.59	2.63
300_07	0	0.59	0.08	0.00
300_08	8	8.00	8.00	8.00
300_09	7	7.46	7.18	7.05
300_10	21	22.60	22.25	21.36
400_01	1	2.52	2.55	1.36
400_02	15	17.37	17.46	16.85
400_03	9	9.91	10.08	10.13
400_04	19	19.01	19.01	19.00
400_05	0	0.01	0.02	0.00
400_06	0	0.33	0.10	0.00
400_07	4	5.44	5.58	4.11
400_08	4	5.30	5.24	4.04
400_09	5	7.63	7.31	6.69
400_10	0	0.95	0.89	0.00

In further numerical experiments, a local search procedure is added to the two ACS algorithms and the GA with NCPX crossover. In the ACS algorithms the local search procedure is applied to the best solution found at each cycle as well as to best overall solution. In the GA, the local search is applied during the 250 first generations to the best solution found if this solution is improved with a probability of 19% as well as to the best overall solution. Table 4 reports the results obtained by the three modified algorithms for SET2 and SET3. First, we observed that the performance of the GA with NCPX crossover combined with local search is globally improved on the two sets. By comparing the three modified algorithms on SET2, one notes that ACS-2D with local search and GA with local search always obtain the best known solution except for instance 10_93. The ACS-3D does not attain the best known solution for instances 10_93 and 16_81 but the deviation from the best known solution for problem 16_81 is negligible. However, the GA approach clearly outperforms the two ACS algorithms on problem 10-93. When we look at the result on SET3, in comparing ACS-2D with local search and NCPX with local search we see that the GA approach outperforms the first ACS algorithm on 24 problems and is worse in only 1 instance. For the 5 remaining instances the 2 algorithms obtain equal results. By comparing NCPX with local search and ACS-3D with local search, one notes that the GA obtains better results on 21 instances, is equal on 6 and is worse on problems 200_01 and 400_03.

5 Conclusions

This paper described three new crossover operators for the traditional car sequencing problem. The performance of these operators has been tested using three standard benchmarks available on the internet. Computational experiments allow us to determinate that the NCPX operator is the best performer. Moreover, the quality of the results produced by this operator appears to be better than that of well known approaches, mainly in large problem instances. One notes that even if the IBX and UIX operators obtain somewhat poorer overall results than NCPX, their performance is still of interest. In future work, we feel that a combination of the three operators in the same algorithm will probably produce good results, although as yet we have no experimental confirmation.

These results, though encouraging, reveal a certain deficiency in the search intensification using our approach as shown by the results of the addition of a local search procedure. In future work, it would be interesting to investigate the use of more sophisticated hybridization mechanisms. It would also be interesting to apply these approaches to more complex industrial problems.

References

1. Parello, B.D., Kabat, W.C., Wos, L.: Job-shop scheduling using automated reasoning: a case study of the car sequencing problem. Journal of Automated Reasoning. 2 (1986) 1-42.
2. Kis, T.: On the complexity of the car sequencing problem. Operation Research Letters, 32(4) (2004) 331-336.

3. Gagné, C., M. Gravel, and W.L. Price, *Solving real car sequencing problems with ant colony optimization.* European Journal of Operational Research. 174(3) (2005) 1427-1448.

4. Gravel, M., Gagné, C., Price, W.L.: Review and comparison of three methods for the solution of the car sequencing problem. Journal of the Operational Research Society. 56(11) (2005) 1287-1295.

5. Delaval, M.: Séquencement des lignes d'assemblage à modèles mélangés. Thèse de doctorat, Université des Sciences et Technologies de Lille. (1997).

6. Lopez, P., Roubellat, F.: Ordonnancement de la production. Hermès Science Publications. Paris. (2001) 431 pages.

7. Gottlieb, J., Puchta, M., Solnon, C.: A study of greedy, local search and ant colony optimization approches for car sequencing problems. Applications of Evolutionary Computing. Lecture Notes in Computers Science. G.R. Raidl *et al.* (Eds). Springer-Verlag Heidelberg. (2003) 246-257.

8. Davenport, A., Tsang, E.: Solving constraint satisfaction sequencing problems by iterative repair. Proceedings of the First International Conference on the Practical Applications of Constraint Technologies and Logic Programming (PACLP). Practical Applications Company, London England. (1999) 345-357.

9. Gagné C., Gravel M., Morin S., Price W.L.: Impact of the pheromone trail on the performance of ACO algorithms for solving the « car sequencing ». Journal of the Operational Research Society. (2006) in press.

10. Solnon, C.: Des fourmis pour le problème d'ordonnancement de voitures. Actes des Journées Francophones de Programmation par Contraintes (JFPC), (2006) 305-316.

11. Warwick, T., Tsang, E.: Tackling car sequencing problem using a generic genetic algorithm. Evolutionary Computation. 3(3) (1995) 267-298.

12. Terada, J., Vo, H., Joslin, D.: Combining Genetic Algorithms with Squeaky-Wheel Optimization. In: Genetic and Evolutionary Computation COnference (GECCO). Seattle (2006).

13. Michalewicz, Z.: Genetic algorithms + data structures = evolution programs. 3rd Edition. Springer-Verlag, Berlin Heidelberg New York (1996).

14. Potvin, J.-Y.: Genetic Algorithms for the Traveling Salesman Problem. Annals of Operations Research. 63 (1996) 339-370.

15. Ben Hamida, S.: Algorithmes Évolutionnaires: Prise en Compte des Contraintes et Application Réelle. Thèse de doctorat. Université de Paris 11 (2001).

16. Holland, J.: Adaptation in Natural and Artificial Systems. University of Michigan Press Ann Harbor (1975).

17. Goldberg, D.E.: Genetic Algorithms in Search, Optimization, and Machine Learning. Massachusetts Addison-Wesley Reading (1989).

18. Basseur, M.: Conception d'algorithmes coopératifs pour l'optimisation multi-objectifs : Application aux problèmes d'ordonnancement de type flow-shop. Thèse de doctorat. Université des Sciences et Technologies de Lille (2004).

19. Barichard, V.: Approches hybrides pour les problèmes multiobjectifs. Thèse de doctorat. École doctorale d'Angers (2003).

20. Berro, A.: Optimisation multiobjectif et stratégies d'évolution en environnement dynamique. Université des Sciences Sociales Toulouse I (2001).

21. Gravel, M., Price, W.L., Gagné, C.: Scheduling jobs in an Alcan aluminium foundry using a genetic algorithm. International Journal of Production Research, 38(2) (2000) 309-322.

22. Zinflou, A.: Système interactif d'aide à la décision basé sur des algorithmes génétiques pour l'optimisation multi-objectifs, Université du Québec à Chicoutimi. Mémoire de maîtrise (2004) 162 pages.

23. Goldberg, D.E., Lingle, R.: Alleles, Loci and the Traveling Salesman Problem. In Proceedings of the First International Conference on Genetic Algorithms (ICGA'85). J.J. Grefenstette Editor. Carnegie-Mellon University Pittsburgh PA (1985) 154-159.

24. Syswerda, G.: Uniform Crossover in Genetic Algorithm. In Proceedings of the Third International Conference on Genetic Algorithms. Morgan Kaufmann Publishers Inc. (1989) 2-9

25. Gent, I.P., Walsh, T.: CSPLib: a benchmark library for constraints. Research Reports of the APES Group. APES-09 (1999) Available from http://4c.ucc.ie/~tw/csplib/ schedule.html.

Author Index

Lecture Notes in Computer Science

For information about Vols. 1–4333

please contact your bookseller or Springer

Vol. 4383: E. Bin, A. Ziv, S. Ur (Eds.), Hardware and Software, Verification and Testing. XII, 235 pages. 2007.

Vol. 4381: J. Akiyama, W.Y.C. Chen, M. Kano, X. Li, Q. Yu (Eds.), Discrete Geometry, Combinatorics and Graph Theory. XI, 289 pages. 2007.

Vol. 4380: S. Spaccapietra, P. Atzeni, F. Fages, M.-S. Hacid, M. Kifer, J. Mylopoulos, B. Pernici, P. Shvaiko, J. Trujillo, I. Zaihrayeu (Eds.), Journal on Data Semantics VIII. XV, 219 pages. 2007.

Vol. 4378: I. Virbitskaite, A. Voronkov (Eds.), Perspectives of Systems Informatics. XIV, 496 pages. 2007.

Vol. 4377: M. Abe (Ed.), Topics in Cryptology – CT-RSA 2007. XI, 403 pages. 2006.

Vol. 4376: E. Frachtenberg, U. Schwiegelshohn (Eds.), Job Scheduling Strategies for Parallel Processing. VII, 257 pages. 2007.

Vol. 4374: J.F. Peters, A. Skowron, I. Düntsch, J. Grzymała-Busse, E. Orłowska, L. Polkowski (Eds.), Transactions on Rough Sets VI, Part I. XII, 499 pages. 2007.

Vol. 4373: K. Langendoen, T. Voigt (Eds.), Wireless Sensor Networks. XIII, 358 pages. 2007.

Vol. 4372: M. Kaufmann, D. Wagner (Eds.), Graph Drawing. XIV, 454 pages. 2007.

Vol. 4371: K. Inoue, K. Satoh, F. Toni (Eds.), Computational Logic in Multi-Agent Systems. X, 315 pages. 2007. (Sublibrary LNAI).

Vol. 4370: P.P Lévy, B. Le Grand, F. Poulet, M. Soto, L. Darago, L. Toubiana, J.-F. Vibert (Eds.), Pixelization Paradigm. XV, 279 pages. 2007.

Vol. 4369: M. Umeda, A. Wolf, O. Bartenstein, U. Geske, D. Seipel, O. Takata (Eds.), Declarative Programming for Knowledge Management. X, 229 pages. 2006. (Sublibrary LNAI).

Vol. 4368: T. Erlebach, C. Kaklamanis (Eds.), Approximation and Online Algorithms. X, 345 pages. 2007.

Vol. 4367: K. De Bosschere, D. Kaeli, P. Stenström, D. Whalley, T. Ungerer (Eds.), High Performance Embedded Architectures and Compilers. XI, 307 pages. 2007.

Vol. 4366: K. Tuyls, R. Westra, Y. Saeys, A. Nowé (Eds.), Knowledge Discovery and Emergent Complexity in Bioinformatics. IX, 183 pages. 2007. (Sublibrary LNBI).

Vol. 4364: T. Kühne (Ed.), Models in Software Engineering. XI, 332 pages. 2007.

Vol. 4362: J. van Leeuwen, G.F. Italiano, W. van der Hoek, C. Meinel, H. Sack, F. Plášil (Eds.), SOFSEM 2007: Theory and Practice of Computer Science. XXI, 937 pages. 2007.

Vol. 4361: H.J. Hoogeboom, G. Păun, G. Rozenberg, A. Salomaa (Eds.), Membrane Computing. IX, 555 pages. 2006.

Vol. 4360: W. Dubitzky, A. Schuster, P.M.A. Sloot, M. Schroeder, M. Romberg (Eds.), Distributed, High-Performance and Grid Computing in Computational Biology. X, 192 pages. 2007. (Sublibrary LNBI).

Vol. 4358: R. Vidal, A. Heyden, Y. Ma (Eds.), Dynamical Vision. IX, 329 pages. 2007.

Vol. 4357: L. Buttyán, V. Gligor, D. Westhoff (Eds.), Security and Privacy in Ad-Hoc and Sensor Networks. X, 193 pages. 2006.

Vol. 4355: J. Julliand, O. Kouchnarenko (Eds.), B 2007: Formal Specification and Development in B. XIII, 293 pages. 2006.

Vol. 4354: M. Hanus (Ed.), Practical Aspects of Declarative Languages. X, 335 pages. 2006.

Vol. 4353: T. Schwentick, D. Suciu (Eds.), Database Theory – ICDT 2007. XI, 419 pages. 2006.

Vol. 4352: T.-J. Cham, J. Cai, C. Dorai, D. Rajan, T.-S. Chua, L.-T. Chia (Eds.), Advances in Multimedia Modeling, Part II. XVIII, 743 pages. 2006.

Vol. 4351: T.-J. Cham, J. Cai, C. Dorai, D. Rajan, T.-S. Chua, L.-T. Chia (Eds.), Advances in Multimedia Modeling, Part I. XIX, 797 pages. 2006.

Vol. 4349: B. Cook, A. Podelski (Eds.), Verification, Model Checking, and Abstract Interpretation. XI, 395 pages. 2007.

Vol. 4348: S.T. Taft, R.A. Duff, R.L. Brukardt, E. Ploedereder, P. Leroy (Eds.), Ada 2005 Reference Manual. XXII, 765 pages. 2006.

Vol. 4347: J. Lopez (Ed.), Critical Information Infrastructures Security. X, 286 pages. 2006.

Vol. 4346: L. Brim, B. Haverkort, M. Leucker, J. van de Pol (Eds.), Formal Methods: Applications and Technology. X, 363 pages. 2007.

Vol. 4345: N. Maglaveras, I. Chouvarda, V. Koutkias, R. Brause (Eds.), Biological and Medical Data Analysis. XIII, 496 pages. 2006. (Sublibrary LNBI).

Vol. 4344: V. Gruhn, F. Oquendo (Eds.), Software Architecture. X, 245 pages. 2006.

Vol. 4342: H. de Swart, E. Orłowska, G. Schmidt, M. Roubens (Eds.), Theory and Applications of Relational Structures as Knowledge Instruments II. X, 373 pages. 2006. (Sublibrary LNAI).

Vol. 4341: P.Q. Nguyen (Ed.), Progress in Cryptology - VIETCRYPT 2006. XI, 385 pages. 2006.

Vol. 4340: R. Prodan, T. Fahringer, Grid Computing. XXIII, 317 pages. 2007.

Vol. 4339: E. Ayguadé, G. Baumgartner, J. Ramanujam, P. Sadayappan (Eds.), Languages and Compilers for Parallel Computing. XI, 476 pages. 2006.

Vol. 4338: P. Kalra, S. Peleg (Eds.), Computer Vision, Graphics and Image Processing. XV, 965 pages. 2006.

Vol. 4337: S. Arun-Kumar, N. Garg (Eds.), FSTTCS 2006: Foundations of Software Technology and Theoretical Computer Science. XIII, 430 pages. 2006.

Vol. 4336: V.R. Basili, D. Rombach, K. Schneider, B. Kitchenham, D. Pfahl, R.W. Selby, Empirical Software Engineering Issues. XVII, 193 pages. 2007.

Vol. 4335: S.A. Brueckner, S. Hassas, M. Jelasity, D. Yamins (Eds.), Engineering Self-Organising Systems. XII, 212 pages. 2007. (Sublibrary LNAI).

Vol. 4334: B. Beckert, R. Hähnle, P.H. Schmitt (Eds.), Verification of Object-Oriented Software. XXIX, 658 pages. 2007. (Sublibrary LNAI).